Digital Transmission Design and Jitter Analysis

For a complete listing of the *Artech House Telecommunications Library*, turn to the back of this book . . .

Digital Transmission Design and Jitter Analysis

Yoshitaka Takasaki

Artech House
Boston • London

Library of Congress Cataloging-in-Publication Data

Takasaki, Yoshitaka.
  Digital transmission design and jitter analysis / Yoshitaka
Takasaki.
      p.   cm.
  Includes bibliographical references and index.
  ISBN 0-89006-503-9
      1. Digital communications. 2. Signal processing--Digital
techniques. 3. Optical communications. I. Title. II. Title:
Jitter analysis.
  TK5103.7.T35 1991   91-27449
  621.382--dc20   CIP

British Library Cataloguing in Publication Data

Takasaki, Yoshitaka
  Digital transmission design and jitter analysis.
  I. Title
  001.64404

  ISBN 0-89006-503-9

© 1991 Artech House, Inc.
685 Canton Street
Norwood, MA 02062

International Standard Book Number: 0-89006-503-9
Library of Congress Catalog Card Number: 91-27449

10 9 8 7 6 5 4 3 2 1

To my family and parents

Table of Contents

Foreword

I first encountered digital communications as an abstract concept in 1967 when I started work as a brand new engineer at Bell Telephone Laboratories in Holmdel, NJ. I remember giving a great deal of thought at that time to trying to understand the essential differences between digital and analog methods of packaging and transporting information. Since then, I have had numerous opportunities to explain these differences to audiences of all types, and must admit that to this day a simple but effective way of capturing the essence of "digital" continues to elude me. This, of course, does not prevent the word from appearing frequently in the popular press, usually with the implication that to choose anything but "digital" is to remain in an uncertain and inflexible past.

When I have only a few seconds to explain the digital concept, at least with respect to why one should prefer digital things, before the listener wanders off onto other, less technical thoughts, I simply say that digital signals are rugged and analog signals are fragile. This is my current best way of capturing the essence of digital. For those of us who are willing to invest a bit more attention to the subject, we know that an explanation of digital *versus* analog transmission must start with the concept of a signal, which represents, either as a mathematical abstraction or as a physical waveform or variable, the value or sequence of values of something of interest. We must then go on to the subject of signal degradations—noises and distortions—that change the signal to some other signal. We are then ready to begin to approach the essence of digital transmission—the ability to remove noises and distortions by recognizing the original signal within its noisy and distorted counterpart. I once tried to explain digital transmission to a friend of mine with an artistic background by suggesting that it would be easier to find and remove a spot of extra paint from a Piet Mondrian painting than from a Jackson Pollock. However, I think I succeeded only in arousing her suspicion that I was over my head in the arts rather than making an attempt to leverage off her expertise.

Of course, for those of us who make our careers in the emerging world of digital telephony, and in particular those of us who are held accountable for making things that really work, digital is a much richer set of concepts than those to which I have alluded so far. In a real sense, we who design digital equipment are in the business of making things that are easy for others to deploy, maintain, and use. A friend of mine who played a major role in the design and construction of some experimental digital coders, which were used to transmit television over fiber optic links during the 1984 Olympics in Los Angeles, told me one day how the television crews found the digital equipment so reliable and free from the need for adjustment that they would arrive at the site of an event just minutes in advance to hook up the necessary gear. This, of course, illustrates one important attribute of digital things—their complexity is built into the boxes, in such a way that outside the boxes, in the field, they appear simple. My friend obviously had mixed emotions about the chances that the television equipment crews were taking with these hand-made prototypes, but all's well that ends well.

I once heard another famous researcher, who has a good deal of interest in music, remark on how much easier it is to build a digital electronic organ, with frequencies and amplitudes controlled by the ticks of a clock (and derivatives thereof), and without the precise adjustments of the values of resistors, capacitors, and the like. He made that observation when integrated circuits were just emerging from the research laboratory. The role of the engineer is to create the details of how one uses a single reference clock to control the frequencies and amplitudes of all of the sounds of a complex musical instrument.

Another dimension of the concept of digital is the magic of low-cost complexity. By now we have all taken for granted the digital tuning and precise control that we obtain digitally on our radios and televisions—while the cost of this disappears in batch processes, involving ever higher levels of integration of imprecise components that nevertheless perform their desired aggregate role exactly as intended.

Every engineer who has entered the domain of digital transmission systems has quickly come to grips with eye diagrams, clock recovery, and jitter. In fact, learning how to model the dynamic behavior of a sequence of repeaters, each with its own clock recovered from its incoming signal, is an initiation rite for those of us who wish to enter the digital communication club. In my first summer at Bell Telephone Laboratories, I modeled the timing response of a chain of digital repeaters to a step in phase caused by a protection switch, using simple analytical methods and an analog computer for simulation (a series of operational amplifiers mounted on a large board with capacitors and resistors that you could attach to make filters). Today, powerful digital computers are used to simulate more complex clock recovery systems with increasing accuracy and speed.

Just the other day, I realized that for all the discussions these days in the press, which mix the concept of the transport of high quality television in digital form with the somewhat separate concept of a digital television set (i.e., a television set with digital signal processing components inside), the value of digitally representing and storing television frames inside a television set, in terms of its ability to perform picture processing functions easily, is hardly mentioned. I suppose this is just another example of why digital technology is an art form like the paintings mentioned earlier—the more you look at it, the more you discover of the beauty that its creators have put into it.

Stewart D. Personick
March 11, 1991

Preface

The evolution of the deployment of digital transmission began as early as 1962, when the first digital transmission system was introduced to increase the capacity of intracity cable routes. Since then, almost three decades have been devoted for the research and developments of digital transmission technologies. Design theory seems to have become almost complete. However, the existence of occasional unknown degradations in transmission lines are still being suspected. Especially, degradations attributable to the analog half of digital transmission, that is, clock recovery, are not necessarily deemed clarified.

Extensive investigations of the technologies and theory for clock recovery were continued for many years after the deployment of digital transmission systems. A coast-to-coast coaxial system, for example, required about 4000 repeatered links of several hundred megabits per second in 4000 miles, where accumulation of timing jitter could cause significant problems. After the advent of low-loss optical fibers in 1970 and succeeding developments that attained dramatic fiber loss reduction, however, the problem of jitter accumulation seemed to become less important, since the required number of repeaters in a link was expected to be reduced by an order of one to two.

On the other hand, in modern local information networks such as *local area networks* (LANs), several hundred digital repeaters can be connected in tandem. Moreover, regeneration of pulse streams and some types of signal processing are incorporated at respective repeaters in such applications. New theories about jitter analyses will have to be investigated, since such processing regenerators are suspected of changing the behavior of jitter accumulation. Therefore, jitter accumulation is still a great concern to transmission system designers.

We should also note that local networks or subscriber loops are becoming more important in digital transmission design; the cost of local systems has now become dominant due to the drastic cost reduction of trunking lines because of the use of broadband and low-loss optical fibers. The use of metallic cable transmission will also have to be considered when designing inexpensive local lines. As a matter of fact, metallic cable systems such as high-bit-rate digital subscriber lines (HDSL), twisted-pair distributed data interface (TPDDI) and 10BASE-T are already being developed for local network applications.

Based on the above observations, this book tries to address new trends in digital transmission design. Some modifications in design theory are considered in order to deal with the suspected new behavior of jitter mentioned previously. We also place more emphasis on local networks than on trunking lines and, therefore, consider the use of not only optical fibers but metallic cables as well.

This book is intended for readers engaged in the design of digital transmission systems and local information networks. Because it is organized to interpret the contents of professional publications and rearrange the story by using more intelligible language, many helpful redundancies, and minimal mathematics, the book can also be used by a reader who wishes to start a career in the field of digital transmission, as well as second-year graduate students in digital communication. This book should also interest readers involved in research and development because it includes new topics in digital transmission with some open questions. The reader is assumed to have previous knowledge about Fourier transform, circuit analysis, and modulation theory—all desirable for starting a career in transmission system design.

Chapter 1 is designed to serve as an executive summary of this book. We discuss theory and technologies for digital transmission design in Chapter 2, Chapter 3, and Chapter 4. We devote Chapter 5 and Chapter 6 to the investigation of new theories about jitter analysis that can accommodate modern information network designs. We learn about a useful tool for digital transmission design in Chapter 7. Finally, we sum up our study in Chapter 8 by investigating general system design methods as well as particular examples of digital transmission such as metallic cable and fiber-optic systems.

Problems are embedded in the text. Their solutions appear at the end of the book.

The author is grateful for the support, help, advice, and input of many people at Hitachi, at other organizations, and in industry, both domestic and worldwide. He also wishes to thank Dr. Kiichi Yamashita for reading the manuscript and providing valuable comments, Miss Kayoko Suzuki for the text processing of this manuscript and Yoshinori for preparing drawings.

Yoshitaka Takasaki
March 11, 1991

Chapter 1

Introduction to Digital Transmission System Design

This chapter is designed to serve not only as an overview, but also as an executive summary for this book. A nonspecialist reader is not required to understand this chapter completely, but is advised to speculate about the answers to be given later, an approach that has shown to improve efficiency in learning [1].

First, we discuss new trends in digital transmission network development, including trunk as well as local lines. We also give an overview of the "3R" functions of digital repeaters and then proceed to budget design study based on the eye diagram with amplitude and time crosshairs.

Next, we review three important technologies in digital transmission: line coding, waveform shaping, and clock recovery. We will see that these technologies can be related by two functions for clock recovery: pattern and waveform functions. We also discuss the usefulness of Bessel filtering in terms of Nyquist's and the Gibby-Smith criteria.

Hence, we investigate jitter generation and accumulation based on Chapman's model. We learn a method that can predict the behavior of root mean square (rms) jitter by using the results of static jitter measurements. We modify Chapman's model to study the phenomenon of extra-jitter accumulation in the presence of processing delay in each repeater of a chain. We apply the concept of a quasistatic criterion for analyzing extra-accumulation.

We then study computer simulation to observe its efficiency in saving time and expense while checking the reliability of theoretical analyses. We shall see that theory and simulation complement each other well in the design and analyses of digital transmission systems.

Finally, we study some examples of digital transmission design. We adopt a hybrid design method because, while rms designs can be unreliable, worst-case designs may impose excessive requirements. We illustrate the usefulness of a hybrid design combined with the quasistatic analysis mentioned above.

1.1 PRINCIPLES OF DIGITAL TRANSMISSION

Transmission systems can be classified into three types: all-electrical (Figure 1.1(a)), hybrid optoelectrical (Figure 1.1(b)), and all-optical (Figure 1.1(c)). A simplified model of a digital transmission network is shown in Figure 1.1(d). In trunk line applications, fiber optic (hybrid optoelectrical) transmission has proved very efficient in attaining repeaterless transmission. However, metallic cable (all-electrical) ap-

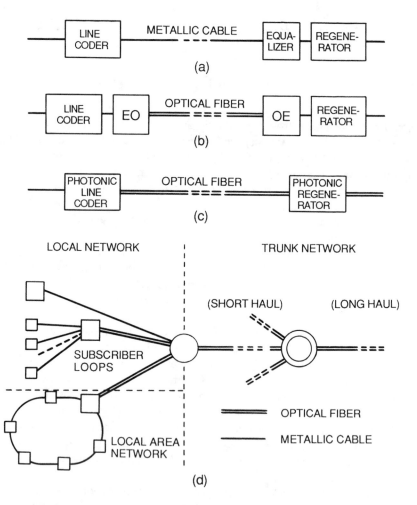

Figure 1.1 Digital transmission lines and network: (a) all-electrical transmission line; (b) hybrid optoelectrical transmission line; (c) all-optical transmission line; and (d) digital transmission network.

proaches are still more appropriate for local systems (Problem 1.1). An extreme example is a twisted-pair-distributed data interface (TPDDI), which provides an economical version of fiber-distributed data interface (FDDI) systems with twisted-pair cables at 100 Mb/s.

Around 75% of total transmission system cost per subscriber is said to be attributable to local systems. Although the all-optical approach is expected to reduce local system cost in future, large-capacity networks, metallic cables are preferred in today's applications. In the near future, fiber optic transmission can play an important role if it offers drastic cost reduction.

Basic components of a digital transmission line are shown in Figure 1.2. At the transmitter, the information pulse stream is applied to the line coder and is converted to assume a pulse format that can facilitate the efficient use of a transmission medium (see Chapter 2). As should be evident, the selection of pulse formats depends on the transmission medium as well as the requirements for transmission system design. Not only optical fibers, but metallic cables such as twisted pair and coaxial cables are considered in this book as transmission media that are adaptable to the new trend of local system development. The regenerator comprises the so-called "3R" functions [2]; reshaping, retiming, and regeneration. The function of the reshaping circuit is usually considered to be the simultaneous reduction of noise and intersymbol interference (ISI). Another important function is the minimization of jitter generation, which is elaborated on in Chapter 5. The retiming circuit is used to recover a high-quality clock component from the incoming pulse stream. Note that the design of the line coder and reshaping circuit has considerable influence on the quality of the recovered clock component. Timing information is carried by waveform transitions in a pulse stream. A pulse format is chosen so the line coder can generate a sufficient number of such transitions per predetermined period. The

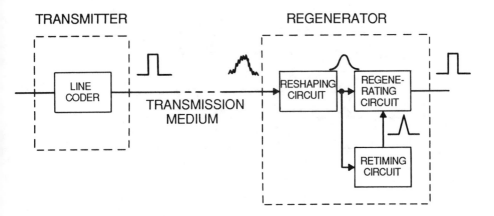

Figure 1.2 Basic components of a digital transmission line.

regeneration circuit makes use of the clock pulse to determine whether each reshaped waveform represents a "mark" or a "space." The reliability of the decision made by the regeneration circuit is affected by impairments in both amplitude and time crosshairs.

The influence of such impairments can be analyzed in terms of the eye diagram shown in Figure 1.3 (solid lines), which is blurred by ISI and noise (dotted lines). The worst-case eye diagram is composed of the lower envelope of a pulse head $\{h\}$, the upper envelope of the preceding pulse tail $\{t\}p$, and the upper envelope of the succeeding pulse tail $\{t\}s$. The area enclosed by these envelopes is the working area for pulse detection. The amplitude and time crosshairs must intersect within this area. We refer to the area as a (worst-case) "eye," and a chart of all eyes as the "eye diagram," which is just an oscilloscopic-type display of a random pulse train synchronized to the pulse repetition rate or frequency.

A simplified design example of a crosshair budget is shown in Table 1.1. A single timeslot duration is expressed as 360° to define time crosshair budgets. Degradations in time crosshair can be converted into equivalent amplitude crosshair deg-

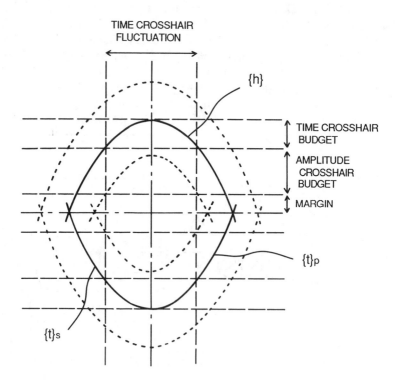

Figure 1.3 Eye diagram.

Table 1.1
Degradations in Eye Diagram (Worst-Case Design Example)

Amplitude Crosshair Degradations	
Noise and/or Crosstalk	25 %
Intersymbol Interference	25 %
Threshold Displacement	10 %
Time Crosshair Degradations	13 % (±60°)
Clock Phase displacement	(±50°)
Alignment Jitter	(±10°)
Margin	27 %

One Timeslot = 360°

radations (Problem 1.2). Both clock phase displacement and alignment jitter cause time crosshair degradations. Alignment jitter is defined as the difference between the jitter of the input pulse stream and that of the recovered clock component (see Section 1.6).

The number of repeaters in tandem can exceed several hundred with such applications as local area networks (LANs) and metropolitan area networks (MANs). Some extra-jitter accumulation may be encountered due to the signal processing required at respective repeaters in such networks. Therefore, both reducing cost, and alleviating jitter accumulation are important problems. We shall study a new theory for analyzing extra-jitter generation and accumulation, as well as the control of jitter in Chapters 5 and 6.

Note that Table 1.1 is organized on the basis of a worst-case criterion, which leads to excessive design requirements. Also, the extra-accumulation of alignment jitter is not taken into account. We study some modification of the design table in Sections 1.8 and 8.1 to solve such problems.

1.2 LINE CODING

The major roles of line codes are the alleviation of waveform distortion and noise, and the improvement of the quality and reliability of clock recovery. We review some representative line codes [3] in this section to help us understand the influence of dc balance and pulse-run length on amplitude and time crosshair degradations (Problem 1.3).

1.2.1 Line Coding for Metallic Cable Systems

The alternate mark inversion (AMI) or bipolar code shown in Figure 1.4 found one of its earliest applications in a commercial twisted-pair cable system [2]. The maximum utilization of a pair cable is obtained if all the pairs within it can be used for transmission. The factor that limits the number of usable pairs is near-end crosstalk (NEXT) (Problem 1.4). AMI line codes are useful for alleviating not only NEXT, but also waveform distortion due to low-frequency cutoff caused by coupling devices, because the energy of the transmitted signals can be concentrated at mid-band ranges. AMI also efficiently alleviates jitter generation because the pulse run length is limited to only one. The timing loss problems associated with the transmission of a long sequence of zeros can be eliminated by applying a zero substitution technique (B6ZS or HDB3), as shown in Figure 1.4.

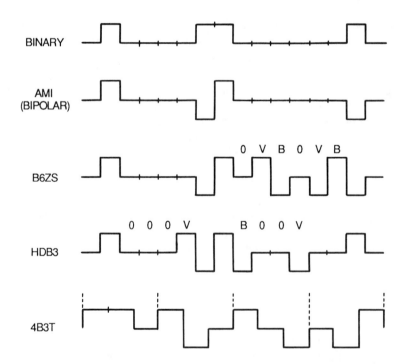

Figure 1.4 Examples of line codes for metallic cable systems.

The bipolar six-zero substitution (B6ZS) code replaces a sequence of consecutive zeros by "0VB0VB" where "0," "V," and "B" indicate a ternary zero, a violation of AMI coding rules, and a mark that obeys AMI rules, respectively. At the receiver, the substituted sequence can be recognized because the violations occur in balanced pairs with two-digit periods between them.

A substitution is made for four consecutive binary zeros with high-density bipolar order three (HDB3). Two sequences, "000V" and "B00V," are used; not balanced in themselves. The choice between the two sequences is made so that the polarity of the successive violations alternates, giving no net dc component. At the receiver, the substitution can be detected as a violation following two zeros.

Coaxial cable differs from paired cable, insofar as it has negligible crosstalk and an accurate attenuation law, with the loss in decibels being closely proportional to the square root of the frequency. Therefore, with coaxial cable systems, the use of high-radix codes to lower the symbol rate, and hence the noise bandwidth, in order to increase the repeater spacing, would seem an attractive approach. One of the typical line codes used for coaxial cable systems is the 4B3T code, shown in Figure 1.4 [4]. We can see that the maximum pulse-run length is increased compared to AMI, which can result in some degradation in jitter performance. We shall study other types of multilevel codes in Section 2.4.

1.2.2 Line Codes for Optical Fiber Systems

Two-level (binary) codes are usually used in optical fiber systems to avoid the influence of nonlinearity in modulating optical sources. Binary line codes can be represented by an *mBnB* family, where m bits of original signal are converted to n bits of coded signal ($m < n$). For lower capacity applications, 1B2B codes have proved useful and economical. For higher capacity applications, in general, choosing a larger m may be preferable. Note, however, that Manchester (biphase) coding has been reported to offer a very attractive alternative to nonreturn-to-zero (NRZ) coding [5], even at a capacity of 250 Mb/s, when an avalanche photodiode is used at the receiver. Let us confine our discussion to line codes with smaller m values ($m = 1–2$) in this section. More general aspects of the *mBnB* family are described in Chapter 2 and elaborated on in [6].

Some examples of a two-level AMI family are shown in Figure 1.5 as binary versions of a ternary AMI format [7]. They are called broadened mark inversion (BMI), coded mark inversion (CMI), and differential mark inversion (DMI). Conversion rules should be self-explanatory. DMI coding is very close to Manchester (biphase) coding (see Section 2.3). Word synchronization is required at the receiver, even with the 1B2B coding, because a single-bit word is converted into a two-bit word. CMI can make the design of word synchronization relatively simple because

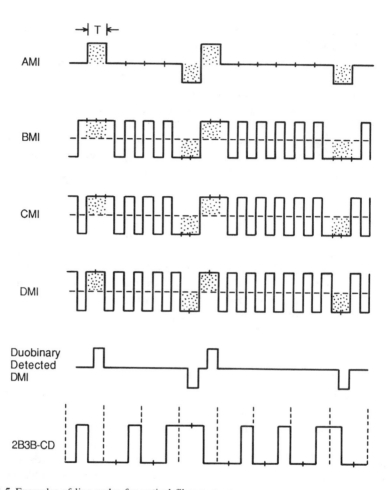

Figure 1.5 Examples of line codes for optical fiber systems.

the phase of word "0" is kept constant. Otherwise, DMI is preferred, owing to its good dc balance and short pulse-run length.

However, applying the technology of correlative coding [8] to DMI not only saves bandwidth, but also eliminates the necessity of word synchronization. We can see from Figure 1.5 that the transmitted two-level sequence with DMI format is duobinary-detected at the receiver, resulting in a three-level sequence with AMI format. Duobinary detection of DMI sequence $d(t)$ can be expressed as $d(t) + d(t + T/2)$, where T is a single timeslot duration. Zero substitution (ZS) can also be applied to a DMI-duobinary system, as explained in Section 2.5.

The required transmission bandwidth can be halved by employing a DMI-duobinary scheme. We can also save bandwidth by using larger m values with the *nBnB* format. Using an odd number for n with dc-balanced codes is undesirable from the standpoint of word synchronization (Section 2.3). A solution is the constant-dc code also shown in Figure 1.5 (2B3B-CD). The implementation of line coding becomes exponentially complicated as the value of m increases. A partitioned-block 8B10B code has been devised for byte-oriented transmission in computer links to alleviate this problem [9].

1.3 WAVEFORM SHAPING

We provide an overview of the conflicts of amplitude *versus* time crosshair designs with waveform shaping. We also study the techniques of equalization for minimizing additional degradations caused by variations in cable responses due to changes in repeater spacing and ambient temperature.

1.3.1 Nyquist's Criterion for Waveform Shaping

Conditions for ISI-free waveform shaping were investigated by Nyquist [10] and extended by Gibby and Smith [11]. Typical examples of Nyquist shaping and the corresponding eye diagrams are shown in Figure 1.6. We can see that reducing bandwidth leads to a reduced time crosshair budget. Also note that a higher degree of phase equalization is required as bandwidth is reduced (Problem 1.5). Also, noise reduction with a smaller roll-off factor β is not necessarily prominent with metallic and optical fiber systems (Problem 1.6). Therefore, 100% roll-off, represented by raised cosine filtering ($\beta = 1$), is usually used.

1.3.2 Minimum Phase Shaping

An exact realization of the raised cosine filter is not physically feasible. Therefore, we must apply some modification to obtain a truncated response (Figure 1.7(a)) or minimum phase response (Figure 1.7(b)); see Appendix A. There, we have to use a Gibby-Smith criterion instead of Nyquist's (Figure 1.7(c)); see Section 3.1. In other words, we need an infinite bandwidth.

Also important is making the waveform as symmetrical as possible to minimize jitter; see Section 5.3. A straightforward approximation of raised cosine filtering can cause significant jitter if phase equalization is not accommodated (Problem 1.7). A practical compromise is to use Bessel or Thompson filtering. This type of filtering closely approximates the raised cosine characteristic in Figure 1.6 ($\beta = 1$) without the use of phase equalization. Some examples of Bessel filter responses are shown

10

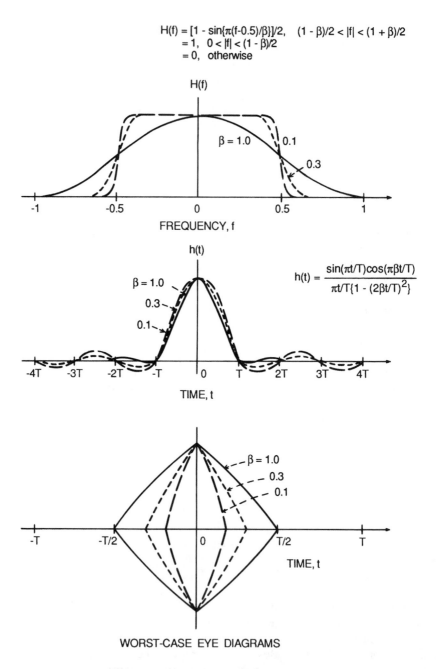

Figure 1.6 Nyquist-type waveform shaping and eye diagrams. *Source*: S.D. Personic, "Receiver Design for Digital Fiber Optic Communication Systems, I," *Bell Syst. Tech. J.*, Vol. 52, July August 1973. p. 857. Reprinted with permission. Copyright © 1973 AT&T.

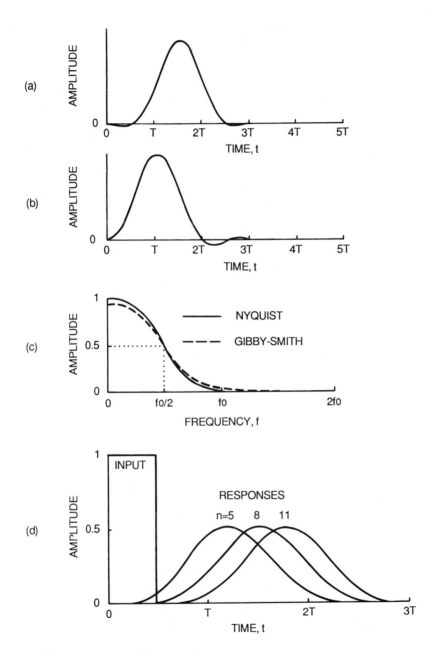

Figure 1.7 Modification of Nyquist's criterion: (a) truncated response; (b) minimum phase response; (c) Nyquist's and Gibby-Smith criteria; and (d) Bessel filter responses.

in Figure 1.7(d). We will see in Section 5.3 that the filter order has to be sufficientl
large to satisfy the alignment jitter requirements in Table 1.1. We can recognize
some asymmetry in the waveform with $n = 5$ in Figure 1.7(d), which may cause a
considerable amount of jitter, as we shall study later. However, waveforms with
$n = 8$ and $n = 11$ may seem to be almost symmetrical, but the former can cause
jitter three times larger than the latter, as we will see in Figure 1.15 below.

General conditions for jitter-free waveform shaping are derived in Section 5.3
It can be shown that a broad-sense symmetrical waveform shown in Figure 1.8(a
with an echo α is free from jitter generation [12]. The waveform comprises a sym
metrical main response with $t = 0$ and its echoes appearing at integer multiples of
pulse period T. The use of this type of waveform adversely affects the reduction of
ISI. Therefore, separate shaping in the timing circuit is needed to avoid this problem

We can analyze the influence of shaping errors on ISI and jitter in terms of
echo theory [13]. Let us study the cases with a single echo, shown in Figure 1.8

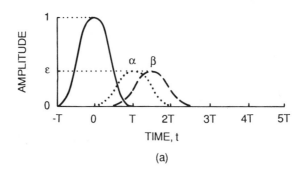

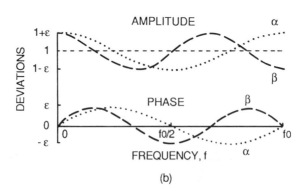

Figure 1.8 Analysis of shaping error in terms of an echo; (a) waveform with an echo, and (b) deviation
of frequency response due to an echo.

to understand the relation of distortions in the time and frequency domains. Let us assume that maximum deviation ε is given in the frequency domain. If a shaping circuit is designed to minimize ISI, an echo tends to appear at the center of a timeslot (echo β in Figure 1.8(a): Problem 1.8), which adversely affects the reduction of jitter. The technique of predistortion can be used to alleviate this problem (Section 5.3).

1.3.3 Equalizing Cable Responses

In metallic cable systems, minimizing equalization errors is important from the standpoint of controlling both amplitude and time crosshair degradations. The so-called "square root of f" characteristics of metallic cables are illustrated in Figure 1.9. Cable loss in decibels is closely proportional to the square root of the frequency. The frequency response $C(f)$ varies depending on repeater spacing and ambient temperature:

$$C(f) = \sqrt{C_0(f)} \; \sqrt{C_0(f)}^u, \quad -1 \leqq u \leqq 1 \tag{1.1}$$

where u represents equivalent cable length variation. Such variations are compensated by using a variable equalizer. The equalization function $E(f)$ can be expressed

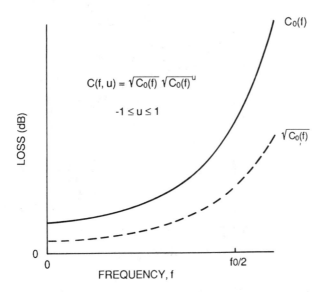

Figure 1.9 Example of metallic cable response.

$$V(f,x) = (x + \sqrt{C_0(f)})/(x\sqrt{C_0(f)} + 1)$$

$$0 \le x \le \infty$$

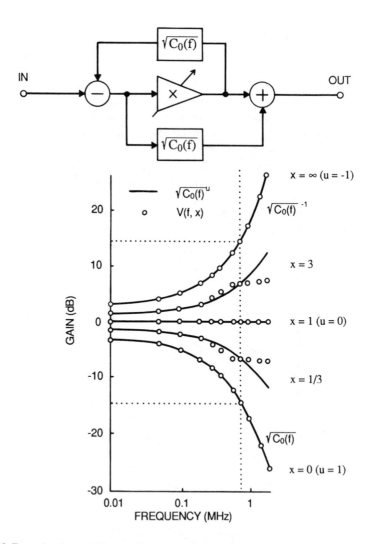

Figure 1.10 Example of a variable equalizer.

as a product of the fixed part $E_f(f)$ and variable part $E_v(f,u)$, where $E_f(f) = 1/\sqrt{C_0(f)}$ and $E_v(f,u) = 1/\sqrt{C_0(f)}^u$. An example of a variable equalizer with a variable function $V(f,x)$ is shown in Figure 1.10 [14], where

$$V(f,x) = [x + \sqrt{C_0(f)}]/[x\sqrt{C_0(f)} + 1], 0 \leqq x \leqq \infty \qquad (1.2)$$

We can see from the figure that $E_v(f,u)$ and $V(f,x)$ coincide at $x = 0$, 1, and ∞ ($u = 1, 0, -1$), where u and x are related by $u = (1 - x)/(1 + x)$. Therefore, we can reduce equalization errors to zero at these three values of u. We also can see that maximum errors are encountered halfway between these values.

This type of variable equalizer utilizes combined feedback and feed-forward. We will also study variable equalizers that incorporate only feed-forward, and are suitable for very-high-speed applications (Section 3.3).

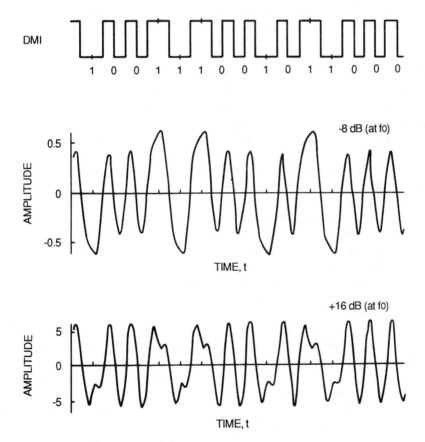

Figure 1.11 Eliminating the use of a variable equalizer by employing DMI coding.

An important consideration is to keep equalization errors as small as possible for all x values in designing $V(f,x)$ to minimize additional jitter due to waveform distortion. Minimizing equalization errors will be studied in Section 3.3.

Note that we can do without the use of variable equalizers at the sacrifice of a doubled bandwidth, that is, by employing DMI-type line coding [15]. Figure 1.11 illustrates that the zero crossings of a received pulse sequence can be kept almost unchanged, regardless of a considerable change in frequency response of the transmission medium. Sliced waveforms can be used to regenerate the pulse sequence. This type of scheme is applied in an IEEE 802.5 token ring LAN.

1.4 CLOCK RECOVERY

Digital transmission comprises not only the transmission of information waveforms, but also that of a clock component. The behavior of a clock component is more complicated than that of information waveforms because it is more analog in nature. The clock component is usually extracted from a received information pulse stream. Figure 1.12(a) shows that a discrete component with a pulse repetition frequency f_0 (dotted line) exists in an information pulse stream, provided an RZ pulse waveform is employed [16]. However, the use of the NRZ format is more common in practical applications, where the clock component cannot be included, as can be seen from Figure 1.12(b). Also note that the waveform is usually reshaped at the receiver in such a way that noise and ISI can be simultaneously minimized. The result of such reshaping can be represented by the raised cosine waveform shown in Figure 1.12(c), which does not include the clock component. Therefore, the use of a discrete component at f_0 can be impractical in most cases.

As is well known, this problem can be solved by applying nonlinear processing to reshaped waveforms. When a pulse stream assumes an NRZ format, we usually apply differentiation before nonlinear processing as shown in Figure 1.13(c). Sinusoidal differentiation (sin$x$), shown in the figure, is useful for suppressing noise in both low- and high-frequency bands. Note that the pulse train after differentiation assumes the AMI format. Therefore, analysis in terms of AMI pulse stream is mainly considered below.

The use of nonlinearity avoids the harmonic and phase distortion problems associated with other schemes involving the linear extraction of timing information from the low-level components received at the pulse repetition frequency f_0. Typical nonlinearities used to this end are full-wave and square-law rectifications (Figure 1.13(d) and (f), respectively). We can obtain very similar results by using either rectification scheme. In the case of full-wave rectification, we can enhance the clock component by clipping the upper half of the pulse (Figure 1.13(e)).

Square-law nonlinearity is convenient for theoretical analyses. We shall show in Section 4.3 that the magnitude (amplitude and phase) of the clock component $X^{(2)}$

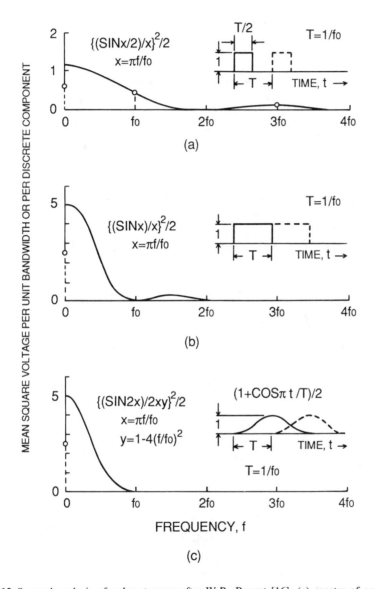

Figure 1.12 Spectral analysis of pulse streams; after W.R. Bennet [16]: (a) spectra of an RZ pulse stream; (b) spectra of an NRZ pulse stream; and (c) spectra of a raised cosine pulse stream.

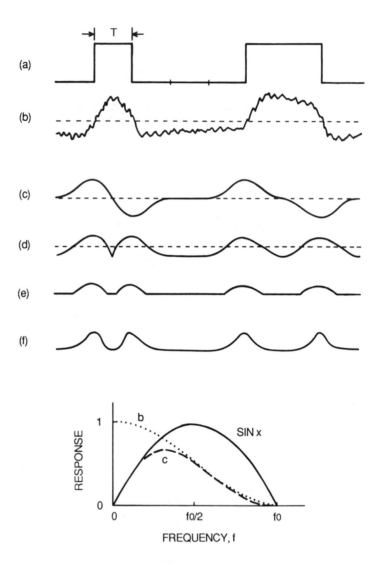

Figure 1.13 Nonlinear processing for clock recovery: (a) transmitted pulse stream; (b) received pulse stream; (c) differentiated pulse stream; (d) full-wave rectified pulse stream; (e) clipped pulse stream; and (f) square-law rectified pulse stream.

(f_0) extracted through square-law nonlinearity can be expressed by an inner product of two functions:

$$X^{(2)}(f_0) = 2 \int_0^{f_0/2} |W(f)|^2 \hat{H}(f) \, df, \ldots \tag{1.3}$$

where $|W(f)|^2$ and $\hat{H}(f)$ are called "pattern function" and "waveform function," respectively [12]. The pattern function depends only on the pattern of a pulse sequence. Actually, $|W(f)|^2$ is equivalent to the power spectrum of an impulse train having the same pattern as that of the pulse stream under consideration. However, the waveform function depends only on the waveform of a single reshaped pulse observed at the input to the square-law device.

Equation (1.3) is very convenient for analyzing the behavior of clock recovery systems, which can be explained by using Figure 1.14. Figure 1.14(a) illustrates amplitude and phase values of the waveform function $\hat{H}(f)$ for a typical waveform after reshaping (i.e., RZ pulse response of a fifth-order Bessel filter shown in Figure 1.7(d)). Figure 1.14(b), (c), and (d) shows pattern functions $|W(f)|^2$(s) for an all-mark sequence with AMI format, an all-mark sequence with straight binary format, and an AMI sequence with a $1/2$ ($=1, 0, 1, 0 \ldots$) pattern, respectively. Let us call them patterns (b), (c), and (d), respectively, below. Because the complex magnitude of the clock component can be given by an inner product of waveform function $\hat{H}(f)$ and pattern function $|W(f)|^2$, we can easily estimate pattern-dependent amplitude and phase variations by checking Figure 1.14. This is studied in the two examples that follow.

Example 1.1

Let us first study an extreme case (i.e., pattern (b) *versus* pattern (c)), which may be impractical, but should be instructive. Such a case can occur in a system where a mark "one" and a space "zero" are represented by $+1$ and -1, respectively. Waveform function $\hat{H}(f)$ is sampled by pattern functions of patterns (b) and (c) in the vicinity of frequencies $f_0/2$ and 0, respectively, in the process of obtaining inner products $X^{(2)}(f_0)$ in (1.3). We can easily see that amplitude and phase variations between these two patterns are approximately 30:1 and almost 50° (one timeslot is defined to be 360°), respectively. Note that variations in the magnitude and phase of the clock component are very large, which can lead to significant impairments in time crosshair.

Example 1.2

Next, let us study patterns (b) *versus* (d). This example is useful for understanding the efficiency of line coding. These two patterns represent all-mark and $1/2$ patterns

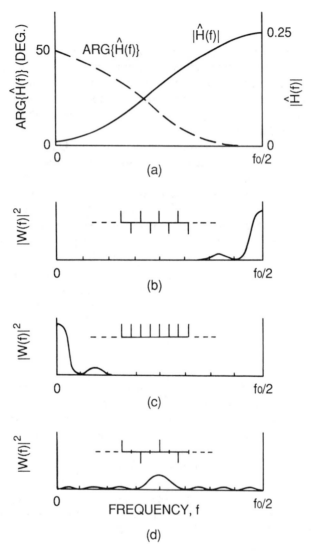

Figure 1.14 Examples of a waveform function and pattern functions; (a) waveform function, an (b–d) pattern functions.

of AMI sequences, respectively. Amplitude and phase variations of the clock component can be estimated to be around 6:1 and 20°, respectively. Note that a remarkable improvement can be attained by the proper selection of line coding as compared with the case in Example 1.1.

1.5 JITTER GENERATION

The classifications of jitter are listed in Table 1.2 (see Tables 5.1 and 6.1 for further jitter classifications). Type A jitter is caused by mistuning, a finite-Q value, and nonzero (NZ) pulsewidth. Asymmetrical waveforms and amplitude-to-phase conversion generate Type B jitter [17], which is what we will mainly study because it becomes predominant compared with other types of jitter, and accumulates in proportion to the number of repeaters in a chain (see Section 6.1).

Table 1.2
Classification of Jitter

Jitter	Accumulation Law
Random Jitter (Noise Dependent)	$\sqrt{N}$
Systematic Jitter (Pattern Dependent)	
Type A	-
Type B	N

N : Number of Repeaters in a Chain

We will not study amplitude-to-phase conversion here, because it is purely related to the design of electronic circuits, and because jitter behavior is considered to be the same as that caused by an asymmetrical waveform. Type B jitter caused by an asymmetrical waveform can be evaluated in terms of the waveform function. We can reduce jitter due to an asymmetrical waveform by employing Bessel filtering with a sufficiently high order, as shown by Figure 1.15(a). The difference between the maximum and minimum phase in a waveform function corresponds to potential worst-case jitter. We can see from the figure that we may not be able to obtain a

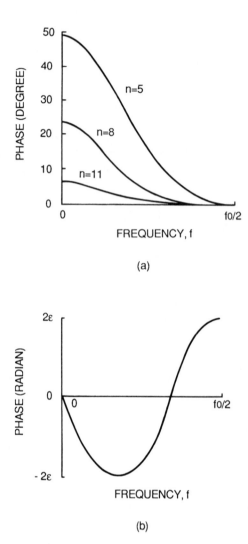

Figure 1.15 Phases of waveform functions; (a) Bessel filtering, and (b) waveform with a single echo

satisfactory result when we use a Bessel filter of the fifth-order. We will learn i Section 5.3 that we can solve this problem by employing a predistortion technique

Another cause of asymmetrical waveforms is shaping errors due to deviation of the shaping circuit response from specified characteristics. Let us study the influence of shaping errors on jitter behavior with the following example.

Example 1.3

The phases of waveform function are shown in Figure 1.15(b), for a waveform having an echo delayed 1.5 timeslots from the main response (echo β in Figure 1.8(a)). Let us choose amplitude of echo $\varepsilon = 0.05$ to cause an ISI of around 10% (Problem 1.9). Then, we can see from Figure 1.15(b) that the maximum jitter is $\pm 6°$ (± 0.1 radian).

We can show that AMI pulse streams in Figure 1.14 (pattern (b) *versus* (d)) cause larger jitter than straight binary ones (pattern (b) *versus* (c)), contrary to Examples 1.1 and 1.2 (Problem 1.10). By checking other patterns (i.e., a 2/4 pattern) however, we may find that the differences in maximum jitter are rather small with these line codes.

1.6 JITTER ACCUMULATION

We have seen how jitter can be generated. Next, let us study the accumulation of jitter in a chain of regenerative repeaters. Two types of jitter must be taken into consideration in designing a digital transmission system: timing jitter and alignment jitter. Timing jitter is also called simply "jitter" or "absolute jitter." As is well known, type B systematic jitter accumulates in proportion to the number of repeaters in a chain [18]. This is because the same jitter $\theta(t)$ is introduced at respective regenerators in a chain, as shown in Figure 1.16(a), Chapman's model (Problem 1.11). Note that, under ordinary circumstances, only the low-frequency components of jitter can accumulate along a repeater chain; higher-frequency components are filtered due to the low-pass nature of timing filters. Therefore, the difference between input jitter and output jitter, called "alignment jitter," is usually rather small. This means that accumulation of alignment jitter is negligible and misalignment of the clock pulse phase from the center of an eye diagram can be kept sufficiently small.

Before proceeding, let us define criteria for evaluating jitter, that is, the static (or transitional) and rms criteria (see Chapter 6, Table 6.1). The static criterion uses step functions to define $\theta(t)$ in Figure 1.16(a). Let us adopt a quasistatic criterion defined by $\theta(t)$, shown in Figure 1.16(b), because the use of the static criterion can lead to excessive requirements in designing crosshair budgets. A square-wave function with a very long period (32,768 timeslots in this example) is used instead of step functions. Jitter accumulation was calculated for a Q value of 100. We can see from the figure that timing jitter accumulates in approximate proportion to the number of repeaters N in a chain; note that the accumulation of alignment jitter is rather small.

A simple and useful method for predicting rms jitter has been developed in [18]. We can use a train of constant phase blocks for $\theta(t)$, as shown in Figure 1.16(c). The phase of each block corresponds to a pulse pattern in the block, which can be

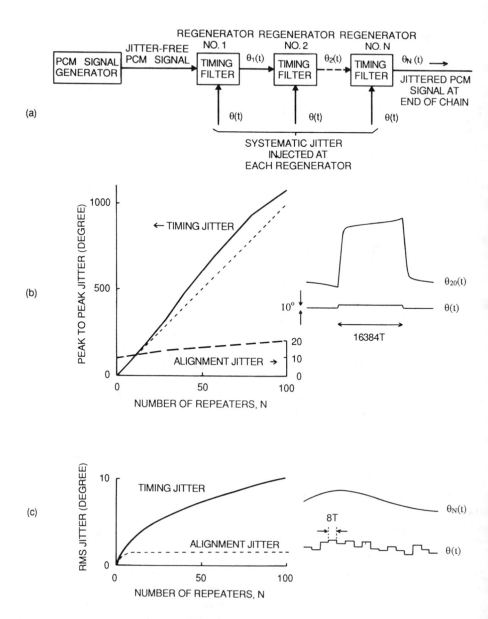

Figure 1.16 Jitter accumulation in a chain of repeaters: (a) Chapman's model for systematic jitter accumulation; *Source*: Byrne, C.J., *et al.*, Systematic Jitter in a Chain of Digital Regenerators, *Bell Syst. Tech. J.*, Vol. 42, No. 6, November 1963, p. 2,681. Reprinted with permission. Copyright © 1963 AT&T; (b) quasistatic analysis of jitter accumulation; and (c) analysis of rms jitter accumulation by using the results of static jitter measurement.

measured by using a repetitive pattern. The length of each block can be as short as eight timeslots (see Chapter 6, Section 6.1). We can see in Figure 1.16(c) that the mean square (ms) value of timing jitter accumulates in proportion to N; also note that the accumulation of alignment jitter is small.

Recently, however, what has been empirically recognized is that a considerable amount of alignment jitter accumulation can be encountered under certain particular conditions [19]. For example, modern information networks require not only regeneration of pulse streams, but also some type of processing at respective repeaters, as shown in Figure 1.17(a). Processing delay M was suspected to cause extra-accumulation of alignment jitter. We will discuss this problem in Section 6.2.

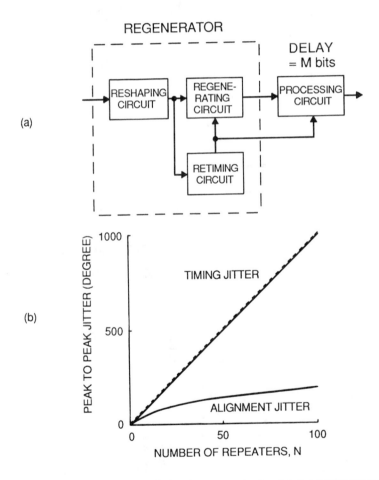

Figure 1.17 Extra-alignment jitter accumulation due to a processing delay; (a) regenerator with a processing delay, and (b) extra alignment jitter accumulation.

Figure 1.17(b) shows the behavior of extra-jitter accumulation in a chain of 100 repeaters. Conditions for calculation are the same as those in Figure 1.16(b), except that M is chosen to be 30 (timeslots) instead of zero. We have to apply some modification to Chapman's model in Figure 1.16(a) because jitter injected at each regenerator is no longer identical, but rather has some delay difference. We can see that timing jitter accumulates exactly in proportion to N, and the accumulation of alignment jitter is around 10 times as large as the case where $M = 0$ (Figure 1.16(b)).

We can qualitatively understand why the extra-accumulation of alignment jitter is encountered by investigating jitter waveforms in Figure 1.18. We can see that

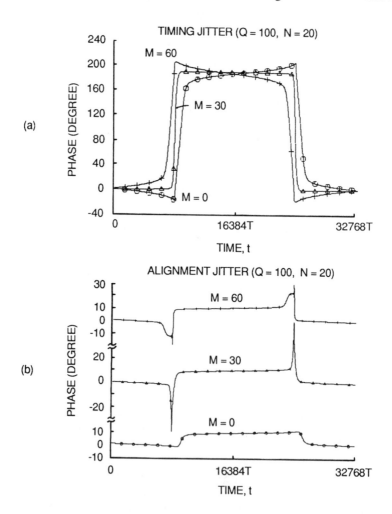

Figure 1.18 Influence of processing delay M on jitter waveform; (a) timing jitter in a chain of 20 repeaters ($N = 20$), and (b) alignment jitter in a chain of 20 repeaters ($N = 20$).

alignment jitter is approximately proportional to the derivative of timing jitter. In this example, $M = 30$ maximizes the transient slopes of timing jitter, and consequently, alignment jitter is maximized.

The dependence of extra-alignment jitter on M values is plotted in Figure 1.19. We can see that the alignment jitter specification of $\pm10°$ in Table 1.1 is no longer appropriate in a system where processing delay is involved in each repeater. We will find in Section 6.2 that the values of M that maximize alignment jitter are almost proportional to Q values.

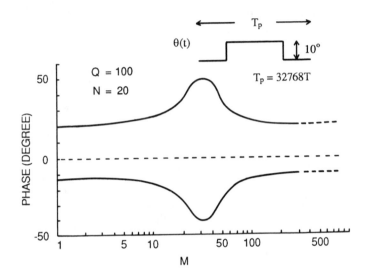

Figure 1.19 Dependence of extra-alignment jitter accumulation on the value of processing delay M (theoretical prediction).

Note that M can assume negative values when narrowband filters such as surface acoustic wave (SAW) devices are used for clock recovery. We can show that extra-accumulation of alignment jitter is not as large as in the case where M is positive (Section 6.2).

1.7 COMPUTER SIMULATION

Checking the reliability of theoretical predictions based on the modified Chapman's model we studied above by using an experimental chain of up to several hundred repeaters or more would be extremely time consuming, and would require tremendous expense. The use of computer simulation should prove very efficient in solving such a problem. The *fast Fourier transform* (FFT) and extremely large memory capacity help to simulate real systems very closely. Details of a computer simulation program are described in Section 7.1.

Let us study the results of computer simulation illustrated in Figure 1.20 to check the theoretical predictions shown in Figure 1.18(b). Two quasistatic pattern combinations, 1/2–8/8 and 1/8–8/8, are considered for representing $\theta(t)$. In general, we can see that good agreement with theory is obtained. Note, however, that the result for 1/8–8/8 exhibits poorer agreement with theory. Negative spikes are shorter than positive ones, as compared with the waveform in Figure 1.18(b) (Problem 1.12). We can also clearly observe some high-frequency fine structures in the waveform of Figure 1.20(b) (Problem 1.13).

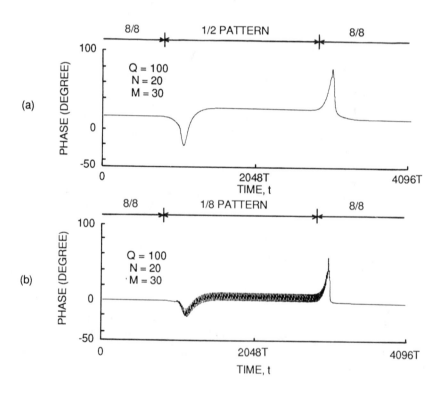

Figure 1.20 Results of computer simulation for extra-alignment jitter accumulation in a chain of 20 repeaters ($N = 20$); (a) alignment jitter for a quasistatic pattern combination of 1/2–8/8, and (b) alignment jitter for a quasistatic pattern combination 1/8–8/8.

The dependence of positive and negative peaks of alignment jitter on processing delay M is plotted in Figure 1.21. We can see that theoretical predictions in Figure 1.19 prove reliable under practical conditions, that is, when alignment jitter accu-

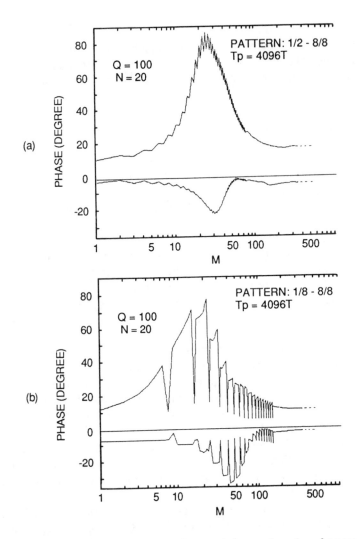

Figure 1.21 Dependence of extra-alignment jitter accumulation on the value of processing delay M (computer simulation); (a) results for a quasistatic pattern combination of 1/2–8/8, and (b) results for a quasistatic pattern combination of 1/8–8/8.

mulation is acceptably small (less than 30° to 40°). The value of M that maximizes extra-jitter accumulation is also predicted by the theory. For this value of M, however, jitter accumulation becomes so large that the theoretical prediction of alignment jitter tends to become rather optimistic, compared with the results of computer simulation shown in Figure 1.21. In addition, we can see that jitter-fine structures make the curves in Figure 1.21 more complicated than those in Figure 1.19. We will discuss this in detail in Section 7.2.2.

Note that pattern period T_p for quasistatic analysis is limited to be less than 4096T with computer simulation due to the upper limits of memory capacity and computation time. Therefore, we can use computer simulation and theoretical analysis to complement each other in the design of digital transmission.

1.8 TRANSMISSION SYSTEM DESIGN

Some examples of digital transmission systems are listed in Table 1.3. Note that three of the sample systems incorporate transmission capacities that are different from transmission rates due to the use of line coding (we study the 10B1C code in Section 2.3). The single-mode fiber system is suitable for trunk line applications; the other systems apply to local applications.

Since the use of a scrambler is assumed in the single-mode fiber system, jitter is defined by rms criterion. Alignment jitter is considered to be almost equal to timing jitter introduced at respective repeaters (that is, 1° rms), since no signal processing is involved in each repeater. The dynamic jitter of 1° empirically corresponds to static jitter of ±10°. Therefore, we can use Table 1.1 for the design of crosshair budget.

Static jitter is specified with the twisted-pair medium, since the use of a scrambler is not considered as it would be in the coaxial cable system (Problem 1.14). A twisted-pair cable of 0.65 ϕ mm can be used to transmit a signal of 150 Mb/s in 200 m, provided the crosstalk problem can be alleviated (Problem 1.15). Neither metallic system specifies alignment jitter, for the same reason the single-mode fiber system does not.

On the other hand, transmitting a signal in a chain of up to 200 repeaters with signal processing functions is considered with the multimode fiber system. There, quasistatic alignment jitter of ±32° has to be accommodated. The use of Table 1.1 is no longer appropriate in such a case.

A hybrid (worst-case combined with rms) criterion, illustrated in Table 1.4, is convenient for alleviating such a problem. The first and second terms of crosshair budget in the table represent worst-case and rms terms, respectively. That is, while the former is added directly, the latter obeys rms-type summation. We can see that alignment jitter of ±32° can be accommodated without sacrificing the budget for margin.

We will also study some design examples of twisted-pair cable and multimode fiber systems in Chapter 8. We will see how the theory and technology we study in this book can be applied to practical systems.

Table 1.3
Examples of Digital Transmission Systems

Transmission Medium	Twisted Pair*	Coaxial Cable*	Multimode Fiber*	Singlemode Fiber*
Transmission Capacity	1.5 Mb/s	100 Mb/s	30 Mb/s	400 Mb/s
Transmission Rate	1.5 Mb/s	75 Mb/s	60 Mb/s	440 Mb/s
Line Code	AMI (6ZS)	4B3T	DMI-Duobinary (8ZS)	10B1C
Repeater Spacing	2 km	2.5 km	3 km	40 km
Jitter per Repeater	$\pm10°$	<1°rms	$\pm10°$ $\pm32°$**	<1°rms (<15°rms)**
Error Rate per Repeater	$<10^{-9}$	$<10^{-11}$	$<10^{-15}$	$<10^{-11}$
Remarks	*0.65 mmø	*1.2/4.4 mm	*Short Wavelength **Alignment Jitter	*Long Wavelength **per 100 repeaters

Table 1.4
Crosshair Budget Design (Hybrid Design Example)

Amplitude Crosshair Degradations		(45 + 8.5%)
Noise and/or Crosstalk		20 + 5%
Intersymbol Interference		20 + 5%
Threshold Displacement		5 + 5%
Time Crosshair Degradations	±{64 + 7.5°}	(15 + 4%)
Clock Phase Displacement		±[32 + 7.5°]
Alignment Jitter : Type A		±[5 + 0°]
Type B		±[27 + 0°]
Margin		30.5%

PROBLEMS

Problem 1.1 Explain why metallic cable systems are still more applicable than fiber optic systems as far as local systems are concerned.

Problem 1.2 Explain how degradations in time crosshair can be converted to equivalent amplitude crosshair degradations.

Problem 1.3 Speculate on the influence of dc balance and pulse-run length on amplitude and time crosshair degradations.

Problem 1.4 Explain what is meant by NEXT.

Problem 1.5 Explain why a higher degree of phase equalization is required as bandwidth shrinks with Nyquist shaping in Figure 1.6.

Problem 1.6 Explain why noise reduction achieved by using a smaller roll-off factor β is not necessarily prominent with metallic and optical fiber systems, compared to wireless systems.

Problem 1.7 Speculate on why a straightforward approximation of raised cosine filtering can cause significant jitter if phase equalization is not accommodated.

Problem 1.8 Explain why an echo tends to appear at the center of a time slot if a shaping circuit is designed to minimize ISI.

Problem 1.9 Explain why echo β in Figure 1.8(a) with $\varepsilon = 0.05$ causes ISI of 10%.

Problem 1.10 Explain why AMI pulse streams in Figure 1.14 (pattern (b) *versus* (d)) cause larger jitter than straight binary ones (pattern (b) *versus* (c)) with the waveform function shown in Figure 1.15(b).

Problem 1.11 It is well known that type B systematic jitter accumulates in proportion to the number of repeaters in a chain [18]. This is because the same jitter $\theta(t)$ is introduced at respective regenerators in a chain as shown in Figure 1.16(a), Chapman's model. Type A jitter is also systematic jitter, but it does not accumulate in a chain of repeaters. Speculate on the reason for this.

Problem 1.12 In the results of computer simulation illustrated in Figure 1.20, we can see that negative spikes are shorter than positive ones, compared to the waveform in Figure 1.18(b). Speculate on the reason for this.

Problem 1.13 We can see some high-frequency fine structures in the waveform of Figure 1.20(b). Speculate on the cause of the fine structure.

Problem 1.14 Explain why the use of a scrambler is considered with the coaxial cable system in Table 1.3.

Problem 1.15 Show that a twisted-pair cable of 0.65 ϕ mm in Table 1.3 can be used to transmit a signal of 150 Mb/s in 200 m, provided the crosstalk problem can be alleviated.

REFERENCES

1. Bynum, M.M., *Speed Learning*, Learn, Inc., 1986.
2. Aaron, M.R., PCM Transmission in Exchange Plant, *Bell Syst. Tech. J.*, Vol. 41, January 1962, pp. 99–141.
3. Special Issue on Coding for Digital Transmission Systems, *Int. J. Electronics*, Vol. 55, No. 1, July 1983.
4. Waters, D.B., Line Codes for Metallic Cable Systems, *Int. J. Electronics*, Vol. 55, No. 1, July 1983, pp. 159–169.
5. Muoi, T.V., Receiver Design for Digital Fiber Optic Transmission Systems Using Manchester (Biphase) Coding, *IEEE Trans. Commun.*, Vol. COM-31, No. 5, May 1983, pp. 608–619.
6. Brooks, R.M., and A. Jessop, Line Coding for Optical Fibre Systems, *Int. J. Electronics*, Vol. 55, No. 1, July 1983, pp. 81–120.
7. Takasaki, Y., Two-Level AMI Line Coding Family for Optical Fibre Systems, *Int. J. Electronics*, Vol. 55, No. 1, July 1983, pp. 121–131.
8. Lender, A., Correlative Digital Communication Techniques, *IEEE Trans. Commun.*, Vol. COM-12, December 1964, pp. 128–135.

9. Widmer, A.X., and P.A. Franaszek, A DC-Balanced, Partitioned-Block, 8B/10B Transmission Code, *IBM J. Res. Develop.* Vol. 27, No. 5, September 1983, pp. 440–451.

10. Nyquist, H., Certain Topics in Telegraph Transmission Theory, *AIEE Trans.*, Vol. 47, April 1928, pp. 617–644.

11. Gibby, R.A., and J.W. Smith, Some Extensions of Nyquist's Telegraph Transmission Theory, *Bell Syst. Tech. J.*, Vol. 44, September 1965, pp. 1487–1510.

12. Takasaki, Y., Timing Extraction in Baseband Pulse Transmission, *IEEE Trans. Commun.*, Vol. COM-20, No. 5, October 1972, pp. 877–884.

13. Wheeler, H.A., The Interpretation of Amplitude and Phase Distortion in Terms of Paired Echoes, *Proc. IRE*, Vol. 27, June 1939, pp. 359–385.

14. Takasaki, Y., *et al.*, Inductorless Variable Equalizers Using Feedback and Feedforward, *IEEE Trans. Circuits and Syst.*, Vol. CAS-23, No. 6, June 1976, pp. 389–394.

15. Takasaki, Y. *et al.*, Digital Transmission Plan for Optical Fibre System, *Proc. Tech. G. Meeting*, IECE, Japan, CS74-86, September 1974, pp. 35–43.

16. Bennett, W.R., Statistics of Regenerative Digital Transmission, *Bell Syst. Tech. J.*, Vol. 37, November 1958, pp. 1501–1542.

17. Manley, J.M., The Generation and Accumulation of Timing Noise in PCM Systems—An Experimental and Theoretical Study, *Bell Syst. Tech. J.*, Vol. 48, March 1969, pp. 541–613.

18. Byrne, C.J. *et al.*, Systematic Jitter in a Chain of Digital Regenerators, *Bell Syst. Tech. J.*, Vol. 42, November 1963, pp. 2679–2714.

19. Takasaki, Y., Alignment Jitter Accumulation in a Chain of Processing Node Regenerators, *Trans. IEICE*, Japan, Vol. E73, No. 10, October 1990, pp. 1712–1716.

20. Hirosaki, B., Systematic Jitter Reduction Effect by Inherent Delay of Acoustic Surface Wave Filter, *Trans. IECE*, Japan, Vol. J59-A, Vol. 7, July 1976, pp. 582–589.

21. Personick, S.D., *Fiber Optics, Technology and Applications*, New York: Plenum Press, 1985, p. 144.

Chapter 2
Line Coding

We shall provide an overview of the primary and secondary requirements for line coding. We study statistical and deterministic analyses of power spectral densities to investigate the influence of line coding on such impairments as jitter, ISI, and crosstalk.

Then, we discuss binary line codes such as dc-balanced codes, CDCs, low-disparity codes, and forced-transient codes—as well as scrambling—to see how they can meet the requirements mentioned above.

We also discuss multilevel codes such as pseudoternary, *mBnT*, and *mBnP*; and correlative (partial response) codes such as duobinary, modified duobinary, and DMI-duobinary, from the same standpoint as above.

Finally, some new approaches to line coding are reviewed: adaptive line codes suitable for simple implementation of asynchronous transmission, to alleviate systematic accumulation of jitter, and multipled block codes, to realize simple clock recovery by means of logical processing.

2.1 PRINCIPLES OF LINE CODING

The words "code" and "coding" are commonly used by different people to mean different things [2]. "Source coding" transforms a message source into a string of binary digits, and "channel coding" converts a string of binary digits to a form suitable for a specific transmission medium. Within the domain of channel coding is a further concentration on what has become known as "line coding." (We shall exclude the topic of error detection and correction here.) Line coding is the conversion to a format that facilitates detection of signals in the presence of transmission impairments. The usual aim is to attain transmission quality, primarily a sufficiently low error rate.

Thus, the primary requirements of a line code are: to minimize vulnerability to ISI and noise; to guarantee stable and reliable clock recovery; to achieve minimization and guarantee with only modest redundancy; to help conditioning of the received signal; and to simplify hardware implementation.

We can use dc-balanced codes, constant-dc codes and low-disparity codes to minimize vulnerability to ISI and noise when low-frequency spectral components of received signal are lost, or significantly attenuated, due to the use of isolating transformers or coupling capacitors. These line codes also help alleviate crosstalk, since they can move spectral components in higher-frequency bands to mid-bands.

Most line codes are useful for enhancing clock recovery, but some of them, such as AMI and duobinary, fail to guarantee stable and reliable clock recovery due to unlimited numbers of zero succession. Either ZS or scrambling can be applied to solve such problems.

Scrambling is most efficient in meeting these first two primary requirements, since no redundancy is required. We have to be careful, however, in selecting line codes, because some of them can entail error propagation. Scrambling can entail error propagation and can even adversely affect the guarantee of clock recovery for particular pulse stream sequences. Correlative codes can save bandwidth without the possibility of error propagation.

The conditioning of received signals, such as *automatic gain control* (AGC), is carried out by detecting the peak value of received pulses and controlling the amplifier gain to keep the peak value constant. Most of the line codes that guarantee clock recovery can be used to detect pulse peaks.

In general, reducing redundancy results in increased hardware scale. Members of the two-level AMI family, which include abundant redundancy, are useful for simplifying hardware implementation. Some types of line codes, such as DMI and Manchester (biphase), are useful for keeping zero crossing from deviating under the influence of waveform distortion: no variable equalizer is required as mentioned in Section 1.3. Also, multiple block codes can be used to considerably simplify clock recovery circuits, as we will study in Section 2.7.

There are also several noteworthy secondary functions which may be incorporated in a line coder, namely; provision for error monitoring, ancillary channels for supervisory functions, and facilitating frame synchronization for multiplexing.

Coding rule violations can be used not only for error monitoring but also for ancillary channels. Randomization of the digit stream by scrambling is required to guarantee reliable synchronization of the transmission frame. Frame formats for *synchronous optical network* (SONET) [3] or *asynchronous transfer mode* (ATM) [4] are useful for multiplexing asynchronous signals. Line codes for adaptive multiplexing that we will investigate in Section 2.6 can prove more economical in some applications.

2.2   SPECTRAL ANALYSES

Analyzing the power spectral density (PSD) of a pulse stream can prove convenient for checking the capability of a line code to minimize vulnerability to ISI and noise, and to guarantee stable clock recovery, as mentioned in Section 2.1.

We can use a line code to suppress high-frequency components of a pulse stream and alleviate crosstalk. It can also suppress low-frequency components to reduce ISI that itself is due to low-frequency cutoff. These features are studied in Example 2.1 below. The possibility of reducing jitter with line coding is studied through the analyses of PSD in Example 2.2.

Example 2.1

PSDs of random impulse streams with AMI format are shown in Figure 2.1, and are supported by [5, 6].

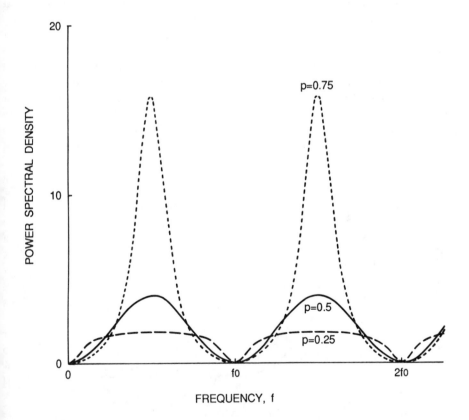

Figure 2.1 Power spectral densities (PSDs) of random impulse streams with AMI format.

$$P_{OAMI}(\omega) = 2(1 - \cos\omega T)/\{1 + 2(2p - 1)\cos\omega T + (2p - 1)^2\} \quad (2.1)$$

where p and T are probability of mark and pulse period, respectively. In actual systems, a pulse shape of RZ or NRZ is employed, with spectral densities $|H_p(\omega)|^2$ given by $[(\sin x)/x]^2$ (see Problem 2.1 for x) as shown in Figure 2.2. PSDs of an RZ pulse stream with AMI format can be obtained by multiplying spectra in Figure 2.1 with those in Figure 2.2, which are shown in Figure 2.3. It is well known that PSDs for a straight binary pulse stream $P_{SB}(\omega)$ with an RZ or NRZ pulse shape are proportional to those in Figure 2.2 [7].

$$P_{SB}(\omega) = \left\{ p(1 - p)/T + p^2 \sum_{n=-\infty}^{\infty} \delta(\omega - n\omega_0)/T^2 \right\} |H_p(\omega)|^2, \quad (2.2)$$

where $\omega_0 = 2\pi/T$, and where δ and n are the delta function and an integer, respectively. Note that while (2.2) includes discrete spectral components represented by δ, no discrete components are included in (2.1). We can understand the efficiency of AMI coding for suppressing low- and high-frequency components through the comparison of spectral shapes in Figures 2.2 and 2.3.

Random pulse streams were assumed in the above analyses. Although they are useful for rms-based analyses, the use of specific pulse patterns can be required for worst-case analyses. Examples of PSD for typical AMI patterns with a sixteen-timeslot length are shown in Figure 2.4. AMI patterns 6/8, 1/2, and 1/4 are defined by the repetition of sequences "1 −1 1 −1 1 −1 0 0", "1 0 −1 0 1 0 −1 0," and "1 0 0 0 −1 0 0 0," respectively. We can see the difference between worst-case design and random sequence design by comparing Figures 2.1 and 2.4.

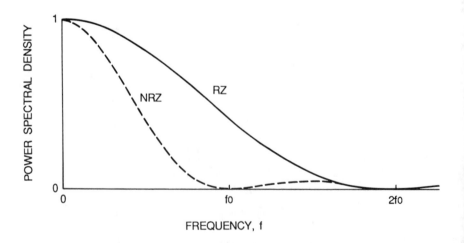

Figure 2.2 Power spectral densities of RZ and NRZ pulses.

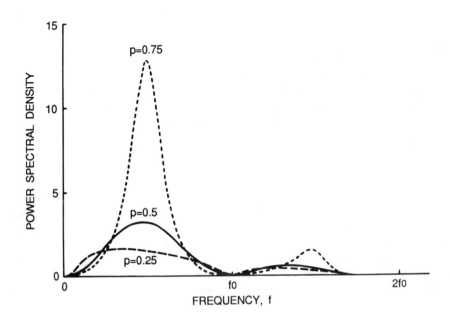

Figure 2.3 PSDs of an RZ pulse stream with AMI format.

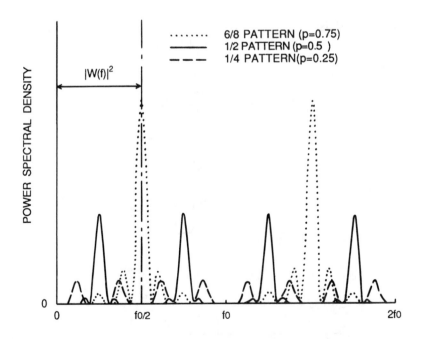

Figure 2.4 Examples of PSDs for typical AMI patterns with a 16-timeslot length.

Example 2.2

As we will discuss in Section 6.1, analyzing jitter in terms of particular patterns is more appropriate than in terms of random sequence. We can use a part of Figure 2.4 to define the pattern function $|W(f)|^2$ (Section 1.4), bounded by a chain line. We can show that a significant amount of jitter can be generated if pattern function $|W(f)|^2$ has spectral components in the vicinity of dc (see Section 1.5). We can see from Figure 2.4 that AMI coding is also efficient in alleviating such jitter generation.

2.3 BINARY LINE CODES

Binary line codes can be represented by the *mBnB* code family, where *m* bits of original signal are converted to *n* bits of coded signal. Codes that belong to the *mBnB* family are suitable for fiber optic transmission and their properties are elaborated in [8]. We can classify the family members, in order of degree of dc balance, as dc balanced (DCB) codes, constant dc (CDC) codes, low disparity (LD) codes and forced transient (FT) codes. Scrambling can be applied with or without these codes in practical systems to meet some of the requirements mentioned in Section 2.1.

2.3.1 DC-Balanced Codes

The two-level AMI family listed in Table 2.1 is the simplest form of *mBnB* codes. Output words "01" and "10" for BMI and DMI codes are chosen to maximize and

Table 2.1
1B2B DC-Balanced Code Translation Table

Input Word	Output Words			
	Manchester (Biphase)	BMI, DMI	CMI 1	CMI 2
0	0 1	01 10	10	01
1	1 0	11 00	11 00	11 00

minimize the run-length of mark and space, respectively. Also, output words "11" and "00" are selected alternately to balance the number of marks and spaces in a pulse stream.

Figure 2.5 compares PSDs for two typical sequences from the two-level AMI family. It is seen that consecutive marks or spaces generate lower-frequency components. Note that a mark and space are represented by $+1$ and -1, respectively, in calculating spectra.

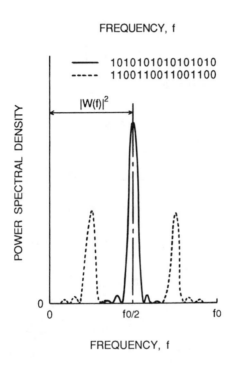

Figure 2.5 PSDs for two typical sequences from a two-level AMI family.

The influence of consecutive marks and spaces is more evident with examples shown in Figure 2.6, where CMI (solid line) and DMI (dotted line) are compared. It is seen that DMI coding is preferable from the standpoint of suppressing lower-frequency components.

Table 2.2 explains why the value of m in $mB(m + 1)B$ codes has to assume an odd number. Output words are selected alternately to balance a pulse stream. It is seen that word synchronization cannot be established, since all possible words are used for carrying information and, therefore, none are available for checking out-of-frame conditions.

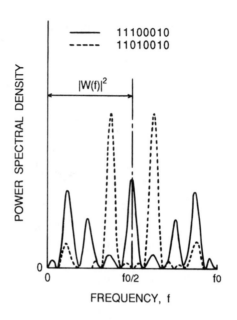

Figure 2.6 Comparison of PSDs for pulse streams having CMI and DMI formats.

Table 2.2
2B3B DC-Balanced Code Translation Table

Input Word	Output Words	
0 0	0 0 1	1 1 0
0 1	0 1 1	1 0 0
1 0	1 0 1	0 1 0
1 1	1 1 1	0 0 0

On the other hand, the 3B4B code listed in Table 2.3 uses 10 out of 16 available words for carrying information. Therefore, six words can be used as violations to detect out-of-word synchronization.

Table 2.3
3B4B DC-Balanced Code Translation Table

Input Word	Output Words	
0 0 0	0 0 0 1	1 1 1 0
0 0 1	0 0 1 1	
0 1 0	0 1 0 1	
0 1 1	0 1 1 0	
1 0 0	1 0 0 1	
1 0 1	1 0 1 0	
1 1 0	1 1 0 0	
1 1 1	1 1 1 1	0 0 0 0

In data communication, information is transmitted in the form of packets with defined field structures for addresses, information, and communication and error control. The number of bits in each of these fields and in the entire packet is generally a multiple of eight. Buffers and relevant interfaces are also byte-oriented. In such systems, an eight-bit code is readily implemented with lower-speed logic on the parallel side of the system. Otherwise attractive binary codes, such as the 5B/6B code, do not readily mesh with a byte-oriented structure.

With these considerations in mind, one would first tend to examine the possibility of an 8B/9B code when looking for code efficiency. However, to realize such a code, implementation and performance parameters other than efficiency must be compromised to a degree too excessive when compared with results we can obtain with less efficient codes. A simple implementation of an 8B/10B code has been reported in [9] for solving such problems, that implementation was accomplished by partitioning the coder into 5B/6B and 3B/4B subordinate coders.

Table 2.4
2B3B Constant-dc Code Translation Table

Input Word	Output Words	
0 0	0 0 1	
0 1	0 1 0	
1 0	1 0 0	
1 1	1 1 0	0 0 0

2.3.2 CDC Codes

CDC line codes can be used to solve the problem with dc-balanced codes as seen in Table 2.4. Five out of eight available words are used to keep the average ratio of mark to space constant (i.e., 1:2). This type of code is suitable for optical transmission systems, since it can keep average optical power output lower than dc-balanced code. Also, the use of an unbalanced threshold can improve bit error rate when an avalanche photo diode is used at the receiver [8].

2.3.3 Low-Disparity Codes

An example of 4B5B low-disparity code is shown in Table 2.5. The number of marks or spaces in a five-digit block is chosen to be two or three. A simple line coder and decoder can be realized at the sacrifice of a small amount of ISI due to low-frequency cutoff (Problem 2.2). Spectral analyses plotted in Figure 2.7 reveals that the worst-case mark succession of six generate spectra at considerably low frequencies. Comparison with Figures 2.5 and 2.6 helps explain the influence of mark or space succession on the generation of low-frequency components.

Table 2.5
4B5B Low-Disparity Code Translation Table

Input Word	Output Words
0 0 0 0	0 0 1 1 1
0 0 0 1	0 0 0 1 1
0 0 1 0	0 0 1 0 1
0 0 1 1	0 0 1 1 0
0 1 0 0	0 1 0 0 1
0 1 0 1	0 1 0 1 0
0 1 1 0	0 1 1 0 0
0 1 1 1	0 1 1 1 0
1 0 0 0	1 0 0 0 1
1 0 0 1	1 0 0 1 0
1 0 1 0	1 0 1 0 0
1 0 1 1	1 0 1 1 0
1 1 0 0	1 1 0 0 0
1 1 0 1	1 1 0 1 0
1 1 1 0	1 1 1 0 0
1 1 1 1	1 1 0 0 1

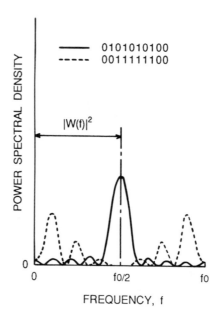

Figure 2.7 Generation of low-frequency components due to a worst-case mark succession with a 4B5F low-disparity code.

2.3.4 Forced Transient Codes

The primary requirement of guaranteeing stable and reliable clock recovery (Section 2.1) can be satisfied by forcing transients with a sufficient frequency, since the timing information is carried by transient waveforms in a pulse stream. An example of transient forcing by using redundant timeslots is shown in Figure 2.8. A complementary symbol (c or $\bar{c}$) for forcing a transient is applied every m timeslots ($m = 8$). This type of line code is called $mB1C$ code [10]. The use of dc restoration is assumed for alleviating ISI degradation due to low-frequency cutoff.

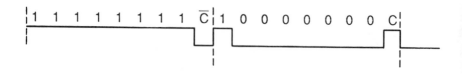

Figure 2.8 Transient forcing by using redundant timeslots.

2.3.5 Scrambling

Scrambling can be used to statistically meet the primary requirements mentioned in Sec. 2.1. A basic scrambler and descrambler of the self-synchronizing type [11] are shown in Figure 2.9. In addition to these basic circuits, monitoring logics are required to avoid desynchronization. Although no redundancy is required with this type of scrambler, transmission errors are multiplied by a factor of w, where w is the number of taps for logical sum in the basic descrambler.

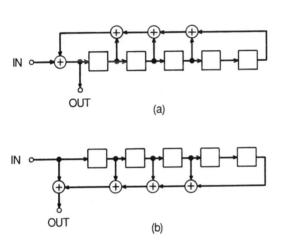

Figure 2.9 Basic scrambler and descrambler of the self-synchronizing type.

External (frame) synchronizing-type scramblers can solve the above problems. However, a frame generator and framer are required at the transmitter and receiver, respectively, for identifying the start and end of a scrambling pattern.

Scrambling can also be useful for alleviating the problem of false word synchronization with block-coded line codes. Since scrambling is statistical in nature and also uses no redundancy, it may not apply well to systems that require extremely high reliability. In such cases, the use of a line code that requires no scrambling, or one that requires no word synchronization, is advised.

2.4 MULTILEVEL CODES

In general, multilevel line codes are expected to be efficient with such high-quality transmission media as coaxial cables. However, such lower-radix codes like pseudoternary codes are also useful with twisted-pair cables.

2.4.1 Pseudoternary Codes

One of the most typical pseudoternary codes is AMI code. It has proved efficient mainly with twisted-pair cable systems. Since it cannot meet the primary requirements, several types of zero-substitution techniques, such as B6ZS and HDB3 (see Section 1.2), have been developed to guarantee transitions in a pulse stream with a sufficient frequency.

Another type of pseudoternary code was proposed for application in a coaxial cable system [12]. In this code, the binary sequence to be transmitted is framed into pairs and translated into ternary format in accordance with Table 2.6. Hence, it is called *paired selected ternary* (PST). There are two modes in the PST code, and they change after each occurrence of either a 10 or a 01 binary pair. Six of the nine possible ternary symbol pairs are used to represent the four possible binary symbol pairs. The remaining three unused ternary symbol pairs are used for framing the pairs at the receiver. Selecting PST or AMI depends on applications (Problem 2.3).

Table 2.6
Paired Selected Ternary Code Translation Table (after I. Dorros, *et al.* [12])

Input Word	Output Words	
0 0	− +	
0 1	0 +	0 −
1 0	+ 0	− 0
1 1	+ −	

Change mode after each 10 or 01

2.4.2 *mBnT* Codes

Although the codes based on formats like AMI and PST require simple code converters, their efficiency is relatively low—63%. Now let us study more efficient codes with *mBnT* formats. The codes representative of this class are 4B3T and 6B4T, with efficiencies of 84% and 95%, respectively.

The code with a 4B3T format, derived by Waters [13], is shown in Table 2.7. There are 27 (3^3) possible combinations of three ternary digits. We can allocate six binary four-bit blocks to the six zero-disparity words. The remaining ten are translated into two alphabets (see Table 2.7.). The disparity is bounded between -3 and $+2$.

Table 2.7
4B3T Code Translation Table (after D. B. Waters [13])

Input Word	Output Words	
0 0 0 0	+ 0 −	
0 0 0 1	− + 0	
0 0 1 0	0 − +	
0 0 1 1	+ − 0	
0 1 0 0	+ + 0	− − 0
0 1 0 1	0 + +	0 − −
0 1 1 0	+ 0 +	− 0 −
0 1 1 1	+ + +	− − −
1 0 0 0	+ + −	− − +
1 0 0 1	− + +	+ − −
1 0 1 0	+ − +	− + −
1 0 1 1	+ 0 0	− 0 0
1 1 0 0	0 + 0	0 − 0
1 1 0 1	0 0 +	0 0 −
1 1 1 0	0 + −	
1 1 1 1	− 0 +	

The average error extension factor is 2.0

We can attain a narrower disparity bound of -2 to $+1$ by using a three-alphabet code, called MS43. The low-frequency spectral components can be reduced, compared with 4B3T, making them more tolerant to ac coupling [14, 15].

A low-disparity code with a 6B4T format can attain higher efficiency than 4B3T codes [16] at the sacrifice of unbounded total disparity. However, it can be shown that this does not significantly affect the transmission performance, provided that a scrambler is used.

2.4.3 *mBnP* Codes

More ambitious multilevel codes with a 6B3P format and two alphabets were proposed for coaxial cable transmission [17]. This five-level code entails a maximum
disparity of six. The disparity boundary can be improved to four by employing four
alphabets [18].

  Many problems are left unsolved with this type of code. For example, including
at least a single pulse with the maximum amplitude (+2 or −2) in a block is desirable
from the viewpoint of detecting the pulse peak value for AGC. It can be shown that
a block length of at least four is required for satisfying this condition. Also, selecting
a coding rule that can minimize amplitude variation of the clock component is important in designing a five-level system [19]. Advances in recent signal processing
technologies are expected to make solving such problems possible in the near future.

2.5 CORRELATIVE (PARTIAL RESPONSE) CODES

Correlative transmission codes proposed by Lender [20] can be used to reduce ISI—
the first primary requirement—when the available bandwidth for transmission is not
large enough. They are also capable of monitoring transmission errors. Correlative
codes are attractive not only with metallic cable systems but also with fiber optic
transmission, since two-level pulses can be used at the transmitter, resulting in multilevel waveforms at the receiver.

  According to Nyquist's theory [21], we can transmit a binary signal at the rate
of T within a bandwidth of $f_0/2$ ($f_0 = 1/T$), by using a $(\sin x)/x$ waveform shown
in Figure 2.10(a). (The dotted waveform is used only to show pulse repetition rate

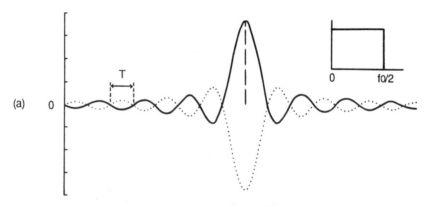

(a)

Figure 2.10 Principles of correlative transmission: (a) $(\sin x)/x$ waveform with oscillating tails; (b)
duobinary waveform with reduced oscillating tails; (c) modified duobinary waveform with
reduced oscillating tails; and (d) minimized oscillatory tails by means of raised cosine
filtering.

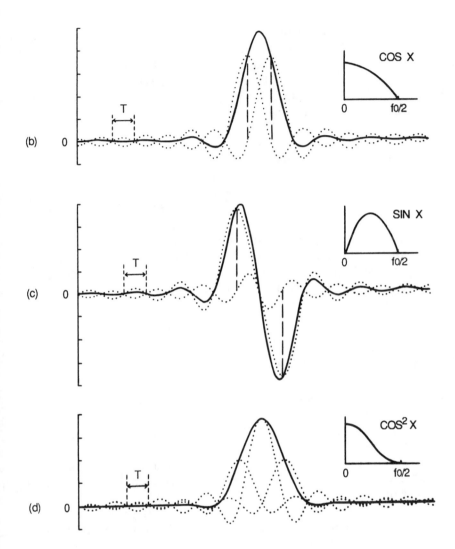

Figure 2.10 Continued.

T.) In this ultimate case, however, no time crosshair budget is available, as is evident in the eye diagrams we studied in Figure 1.6. Oscillating tails of the waveform significantly reduce time crosshair margin.

Transmitting two consecutive pulses instead of a single pulse to represent a symbol can alleviate this problem, since we can make oscillating tails quite small (Figure 2.10(b)). Such a waveform has been termed duobinary [22]. (Note that this

waveform satisfies Nyquist's second criterion [21].) A duobinary waveform can b
obtained as an impulse response of the cosine filter shown in the figure. Note als
that a pulse stream is no longer binary, but assumes three levels.

Another solution, shown in Figure 2.10(c), has been termed modified duobi
nary [23]. This is equivalent to transmitting 1 0 −1 pulses. The waveform can als
be obtained as an impulse response of a sine filter. A unipolar binary pulse strean
is converted into a dc-free three-level sequence.

A third solution, shown in Figure 2.10(d), incorporates a 1/2, 1, 1/2 sequenc
to generate a waveform with the smallest of oscillatory tails. This solution result
in a five-level pulse stream. Note that this waveform can be deemed an impuls
response of a raised cosine filter. As a consequence, the same transmission capacit
can be obtained by using this waveform to transmit a four-level sequence with
pulse rate of $2T$.

The use of the first and second solutions mentioned above will be considere
for improving the efficiencies of transmission. One problem in using waveforms i
Figure 2.10(b) and (c) is the propagation of errors in recovering original signals a
the receiver, solutions to which are described in the following sections. We will als
consider applications to optical fiber systems.

2.5.1 Duobinary Transmission

A simple example in Figure 2.11 explains the problem of error propagation whe
the duobinary waveform in Figure 2.10(b) is adopted for transmission. Transmittin
original pulse stream (a) in Figure 2.11 with a duobinary waveform is equivalent t
delaying the stream by one timeslot (stream (b)) and adding the delayed stream t
the original. This results in a three-level stream (c). For clarity, the duobinary wave
form is shown in an impulse form, rather than in the shaped form that actually ap
pears after filtering.

Recovery of the original sequence can be accomplished simply by detecting
the top level of the three levels, noting that it was formed by an algebraic additio
of the present and previous pulse, as in Figure 2.11 (c). The present pulse to b
detected corresponds to binary "1" as does the previous one. Thus the top level a
the sampling instant can be interpreted as binary "1" transmitted. Similarly, the bot
tom level sent can be interpreted as binary "0" since it was formed by the absenc
of two pulses. The center level was formed by a single pulse and no pulse, corre
sponding to 10 or 01, so the transmitted bit can be interpreted as either "1" or "0
depending on the previous bit. Thus, the interpretation of the center level cannot b
made without researching its history. A wrong decision at the center level will resul
in error propagation; that, of course, will stop as soon as one of the extreme level
is reached.

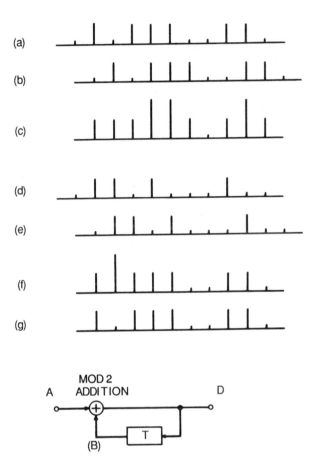

Figure 2.11 Principles of duobinary transmission: (a) original pulse stream; (b) pulse stream (a) delayed by one timeslot; (c) pulse stream obtained by superposing streams (a) and (b) (duobinary waveform); (d) response of the digital feedback circuit to pulse stream (a); (e) pulse stream (d) delayed by one timeslot; (f) pulse stream obtained by superposing streams (d) and (e) (duobinary waveform); and (g) pulse stream decoded from stream (f).

To remedy this situation, the encoding method shown at the bottom of Figure 2.11 was proposed [22]. The binary data input first passes through a digital feedback circuit between points A and D. Then pulse stream (d) can be obtained. Transmitting stream (d) with the duobinary waveform results in stream (f). In spite of the correlation properties of the waveform, it is possible now to make the bit decisions without researching its history. Consequently, error propagation is eliminated. The extreme levels now correspond to binary "0," and the center level to binary "1" (pulse stream (g)).

The duobinary waveform has some interesting properties. A transition between two successive extreme levels is not possible—transitions occur only between adjacent levels. This considerably reduces ISI. The number of bits at the center level determines the polarity of successive extreme levels. If the number is even, the polarities are identical; otherwise, the polarities are opposite. Thus, a predetermined pattern is established and any violations due to noise or other impairments can be monitored.

2.5.2 Modified Duobinary Transmission

Encoding and recovery of the modified duobinary system (Figure 2.10(c)) is illustrated in Figure 2.12. Unlike the duobinary signal, the modified duobinary waveform has no dc component and is therefore easily adaptable to dc-free transmission. However, the ISI at the transition points is greater because a transition between any two levels is permitted.

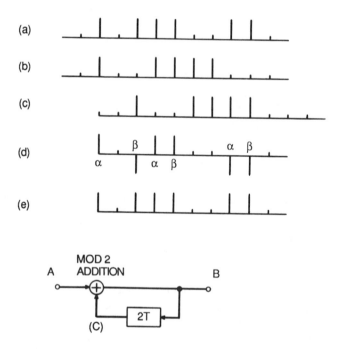

Figure 2.12 Principles of modified duobinary transmission: (a) original pulse stream; (b) response of the digital feedback circuit to pulse stream (a); (c) pulse stream (b) delayed by two timeslots; (d) pulse stream obtained by subtracting stream (c) from stream (b); and (e) pulse stream decoded from stream (d).

As would be expected, due to its correlation properties, the modified duobinary signal also follows a predetermined set of rules. These rules can be deduced by grouping all successive binary 1s in pairs and assigning the pair label to each binary "1" as shown in Figure 2.12(d). A binary "1" bearing α in a pair of two successive binary 1s always has the opposite polarity in modified duobinary relative to the previous binary "1"—which, of course, carries label β. The polarity of the binary "1" that has label β, relative to the previous binary "1" bearing α, is governed by the set of odd and even rules, as in the duobinary case, thus, if the number of intervening 0s between a pair of 1s bearing α and β is even, their polarities are identical, otherwise they are opposite. This statement can be verified for the modified duobinary waveform in Figure 2.12(d).

2.5.3 Modified Duobinary Class II

By checking Figure 2.10, we see that using pulse peak for AGC could be difficult with duobinary systems (Problem 2.4) unless scrambling is used. Pulse peaks can be kept constant at the sacrifice of a doubled bandwidth as is illustrated in Figure 2.13 [24]. Delay times in encoding and decoding and halved, compared with those in Figure 2.12, and result in a received pulse stream with the AMI format (Figure 2.13(d)). A class II waveform is similar to the one in Figure 2.10(c) except that T is halved. Also $f_0/2$ is doubled with sin x curve in the figure. Note that time crosshair budget can be increased by further applying $\cos^2 x$ filtering to make the total response a composite $\sin x \cos^2 x$ characteristic (Problem 2.5).

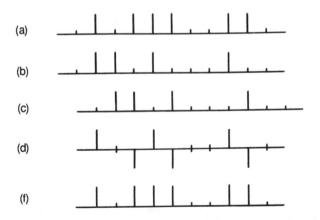

Figure 2.13 Principles of modified duobinary class II: (a) original pulse stream; (b) response of a digital feedback circuit to pulse stream (a); (c) pulse stream (b) delayed by one timeslot; (d) pulse stream obtained by subtracting stream (c) from stream (b); and (e) pulse stream decoded from stream (d).

2.5.4 DMI-Duobinary

Applying ZS in a duobinary system is rather difficult. Combining another line coding scheme, such as DMI, is helpful in solving this problem. An example is shown in Figure 2.14 [24–26]. The original pulse stream (a) is converted into a DMI stream (b). A duobinary-detected waveform (d) can be obtained by taking the algebraic sum of stream (b) and delayed stream (c). Subtracting the continuous all-mark pulse from (d) results in a stream with the AMI format (e).

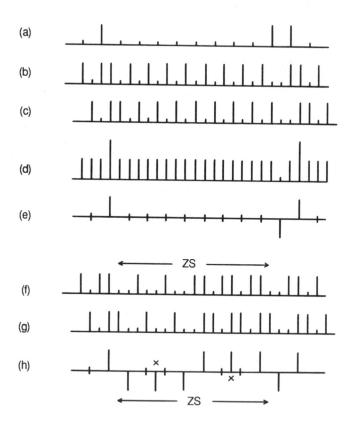

Figure 2.14 Principles of DMI-duobinary transmission with ZS: (a) original pulse stream with eight-zero succession; (b) stream (a) converted to assume DMI format; (c) pulse stream (b) delayed by half a timeslot; (d) pulse stream obtained by superposing streams (b) and (c); (e) pulse stream obtaine by subtracting all-mark pulse from stream (d); (f) pulse stream that includes a ZS pattern with DMI law violations; (g) pulse stream (f) delayed by half a time slot; and (h) pulse stream obtained by superposing streams (f) and (g).

Pulse stream (f) includes a ZS pattern with DMI law violations. This ZS pattern results in a three-level sequence in (h) with AMI law violations and out-of-phase pulses (marked "x"). Since the timing component is cancelled by the "x" pulses, the equivalent number of pulses that can contribute to clock recovery is two $(4 - 2)$ for each ZS pattern. Further details of the DMI-duobinary system are described in Section 8.3.

2.6 ADAPTIVE LINE CODES

It is expected from the discussion in Section 1.6 that systematic accumulation of timing jitter can be alleviated by adopting asynchronous transmission. Figure 2.15 shows examples of simple line codes that can adapt any asynchronous pulse stream to a predetermined line rate [25,27].

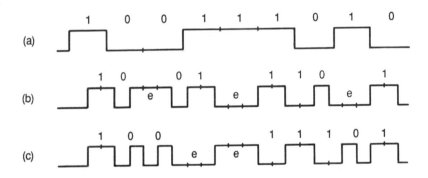

Figure 2.15 Principles of rate adaptive line code: (a) original pulse stream; (b) rate adaptation without dc balance; and (c) rate adaptation with dc balance.

Original symbols "1" and "0" in (a) are converted to "11/00" and "1/0," respectively, as shown in (b). When no information signal is available an empty indication, "111/000," is inserted. Shown in (c) is a dc-free version of this type of code, where "1," "0," and "e" are converted into "11/00," "01/10," and "111000/000111," respectively. Code (b) can enjoy higher efficiency than code (c) provided a scrambler can be used.

2.7 CLOCK-ORIENTED CODES

Line coding is also useful in reducing the cost of a clock recovery circuit. Such a cost can dominate total receiver cost. Examples of such line codes are shown in

Figure 2.16 [28]. Each block in that figure's pulse stream (c) includes only a single information digit, represented by a dotted line. The other three digits are devoted to clocking purposes. Note that very simple logical processing can be applied to recover the clock component. Delaying the pulse stream by two digits (d) and adding it to the original pulse stream (c) results in a recovered clock component as shown in (e). Coding efficiency is 25% in this example.

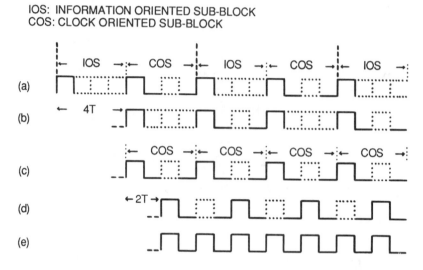

Figure 2.16 Principles of clock oriented code: (a) pulse stream with paired sub-blocks, IOS and COS; (b) pulse stream delayed by four timeslots; (c) pulse stream obtained by the logical multiplication of streams (a) and (b); (d) pulse stream delayed by two timeslots; and (e) clock component recovered by logical addition of streams (c) and (d). *Source:* Takasaki, Y., Multiplexing and Transmission Systems for All-Optical Networks, *ICC '90*, April 16–19, pp. 1,668–1,672, Reprinted with permission. Copyright © 1990 IEEE.

Improving efficiency with a modification of this type of code is illustrated in (a). Each block in the pulse stream shown there is composed of two types of sub-blocks. One is a clock-oriented sub-block (COS), the other, an information-oriented sub-block (IOS). The former is supposed to assume the same format as the one shown in (c). The latter is used to improve coding efficiency. In this example, every IOS includes three information digits and a single digit devoted to clocking. It is difficult to logically recover the clock component from a pulse stream with such combined formats, but it can be shown that the pulse stream can be converted into another stream that comprises only sub-blocks with a COS format. That is, delayig the pulse stream by four time slots and multiplying it by the original pulse train, results in the pulse stream shown in (c), which is composed only of sub-blocks with a COS format. Therefore, the clock component can be extracted by applying the same logical processing to a pulse stream as mentioned above after the conversion.

It should be evident that clock recovery achieved through the logical processing mentioned above can be implemented with a very simple circuit. This type of line code is especially suitable for short distance transmission where degradation of waveform is rather small and signal-to-noise ratio is relatively high.

PROBLEMS

Problem 2.1 Determine x for power spectral density $[(\sin x)/x]^2$ in Figure 2.2 with RZ and NRZ waveforms.

Problem 2.2 Calculate the amount of worst-case ISI generated due to low-frequency cut-off with the 4B5B low-disparity code in Table 2.5.

Problem 2.3 Compare PSD and AMI from the standpoint of practical applications.

Problem 2.4 Show that using pulse peak for AGC can be difficult with a modified duobinary pulse stream when scramblers are not used.

Problem 2.5 Show that time crosshair budget can be increased by applying $\cos^2 x$ filtering to $\sin x$ filtering in Figure 2.10(d).

REFERENCES

1. Special Issue on Coding for Digital Transmission Systems, *Int. J. Electronics*, Vol. 55, No. 1, July 1983.
2. Cattermole, K.W., Principles of Digital Line Coding, *Int. J. Electronics*, Vol. 55, No. 1, July 1983, pp. 3–33.
3. Ballart, R., and Y. C. Ching, SONET: Now It's the Standard Optical Network, *IEEE Commun.*, March 1989, pp. 8–15.
4. Ohta, N., *et al.*, Video Distribution on ATM-Based Optical Ring Networks, *Conf. Rec. ICC '90*, April 1990, pp. 976–980.
5. Aaron, M.R., PCM Transmission in Exchange Plant, *Bell Syst. Tech. J.*, Vol. 41, January 1962, pp. 99–141.
6. Cariolaro, G.L., G. L. Pierobon, and G. P. Tronica, Analysis of Codes and Spectra Calculations, *Int. J. Electronics*, Vol. 55, No. 1, July 1983, pp. 35–79.
7. Bennett, W.R., Statistics of Regenerative Digital Transmission, *Bell Syst. Tech. J.*, Vol. 37, November 1958, pp. 1501–1542.
8. Brooks, R.M., and A. Jessop, Line Coding for Optical Fibre Systems, *Int. J. Electronics*, Vol. 55, No. 1, July 1983, pp. 81–120.
9. Widmer, A.X., and P. A. Franaszek, A DC-Balanced, Partitioned-Block, 8B/10B Transmission Code, *IBM J. Res. Develop.* Vol. 27, No. 5, September 1983, pp. 440–451.
10. Yoshikai, N., K. Katagiri, and T. Ito, MBIC Code and Its Performance in an Optical Communication System, *IEEE Trans. Commun.*, Vol. COM-32, No. 2, February 1984, pp. 163–168.

11. Savage, J.E., Some Simple Self-Synchronizing Digital Scramblers, *Bell Syst. Tech. J.*, Vol. 46, 1967, pp. 449–497.

12. Dorros, I., J. M. Sipress, and F. D. Waldhauer, An Experimental 224 Mb/s Digital Repeatered Line, *Bell Syst. Tech. J.*, Vol. 45, September 1966, pp. 993–1042.

13. Waters, D.B., Line Codes for Metallic Cable Systems, *Int. J. Electronics*, Vol. 55, No. 1, July 1983, pp. 159–169.

14. Franaszek, P.A., Sequence State Coding for Digital Transmission, *Bell Syst. Tech. J.*, Vol. 47, 1968, pp. 143–157.

15. Buchner, J.B., Ternary Line Codes, *Philips Telecommun. Rev.*, Vol. 34, 1976, pp. 72–86.

16. Catchpole, R.J., Efficient Ternary Transmission Codes, *Electron. Lett.*, Vol. 11, 1975, pp. 482–484.

17. Aratani, T., and H. Fukinuki, 800 Mb/s PCM Multilevel Transmission System via Coaxial Cables, *Electrical Commun. Labs. Tech. J.*, Vol. 19, No. 6, 1970, pp. 1181–1198.

18. Miyata, M., unpublished technical memorandum, September 1971.

19. Takasaki, Y., Self-Timing in Digital Multilevel Transmission, *Inst. Electro. Commun. Eng.*, Japan Technical Meeting Report, CS 71-16, June 1971.

20. Lender, A., Correlative Digital Communication Techniques, *IEEE Trans. Commun. Tech.*, Vol. COM-12, December 1964, pp. 128–135.

21. Nyquist, H., Certain Topics in Telegraph Transmission Theory, *AIEE Transactions*, Vol. 47, April 1928, pp. 617–644.

22. Lender, A., The Duobinary Technique for High Speed Data Transmission, *IEEE Trans. Commun. and Elect.*, Vol. 82, May 1963, pp. 214–218.

23. Lender, A., Correlative Level Coding for Binary-Data Transmission, *IEEE Spectrum*, Vol. 54, February 1966, pp. 104–115.

24. Takasaki, Y., M. Tanaka, N. Maeda, K. Yamashita and K. Nagano, Optical Pulse Formats for Fiber Optic Digital Communications, *IEEE Trans. Commun.*, Vol. 24, No. 4, April 1976, pp. 404–413.

25. Takasaki, Y., Two-Level AMI Line Coding Family for Optical Fibre Systems, *Int. J. Elect.*, Vol. 55, No. 1, July 1983, pp. 121–131.

26. Takasaki, Y., and Y. Takahashi, DMI-Duobinary Transmission Experiment in a Chain of Fiber Optic Repeaters, *Conf. Rec. ICC '83*, Boston, June 19–22, 1983, pp. 100–104.

27. Takasaki, Y., *et al.*, Fundamental Properties of Asynchronous Multiplexing for Fiber Optic Adaptive Transmission, *Trans. IECE Japan*, Vol. E64, 1981, pp. 109–114.

28. Takasaki, Y., Multiplexing and Transmission Systems for All-Optical Networks," *Conf. Rec. ICC '90*, April 16–19, 1990, pp. 1668–1672.

Chapter 3
Waveform Shaping

We shall study Nyquist's theory, and extended versions of it, to learn how to minimize ISI. Then we shall investigate criteria for minimizing jitter with the help of the waveform function. We will show that minimizing ISI can result in increased jitter. We shall also study Bessel filtering for compromising ISI and jitter under practical (minimum phase) conditions.

Then, we will study equalizer design for compensation of waveform distortions introduced by transmission cables. We will mainly discuss inductorless realizations and wide-variable-range designs, which require the use of more mathematics than in other parts of this book. Bilinear and biquadratic types of variable equalizers can be realized with the combined use of feedback and feed-forward. Linear and quadratic types of variable equalizers that utilize only feed-forward are also studied for high-speed applications.

Finally, we will study paired echo theory for analyzing the influence of shaping errors on ISI and jitter.

3.1 CRITERIA FOR WAVEFORM SHAPING

3.1.1 Controlling ISI

A criterion for ISI-free waveform shaping investigated by Nyquist [1] is shown in Figure 3.1(a). Curves b and b' are symmetrical about $f_0/2$. We can obtain waveforms without ISI if

$$a(f) + b'(f) = 1 \quad \text{for} \quad 0 \leq f \leq f_0/2 \tag{3.1}$$

In practical systems, however, it is impossible to completely eliminate spectral components above f_0; note that a tail (shown by a dotted line) extends over f_0.

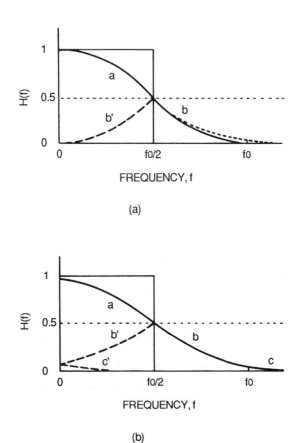

Figure 3.1 Criteria for ISI-free waveform shaping; (a) Nyquist's criterion, and (b) Gibby-Smith criterion.

Extended criteria developed by Gibby and Smith [2] can be applied in such a case. Figure 3.1(b) shows one of the simplest examples. A condition for ISI-free transmission can be given by

$$a(f) + b'(f) + c'(f) = 1, \quad 0 \le f \le f_0/2 \qquad (3.2)$$

where

$$b'(f) = b(f_0 - f) \quad \text{and} \quad c'(f) = c(f - f_0)$$

Two typical waveforms and their spectral responses suggested by Sunde, as examples with zero or negligible ISI [3] are shown in Figure 3.2. Spectral responses

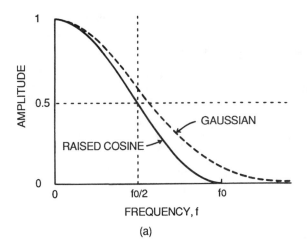

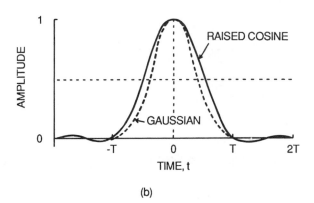

Figure 3.2 Responses with zero or negligible ISI. After E.D. Sunde [3]; (a) frequency domain responses, and (b) time domain responses.

of a Gaussian waveform $H_G(f)$ and a raised cosine waveform $H_R(f)$ are given by

$$H_G(f) = \exp[-\pi(f/f_g)^2] \tag{3.3}$$

$$H_R(f) = (1/2)\{1 + \cos(\pi f/f_r)\}, \quad f \le f_r$$

$$= 0 \qquad\qquad\qquad f > f_r \tag{3.4}$$

It is seen that $H_R(f)$ satisfies a Nyquist criterion when f_r is chosen to be f_0. It can be shown that $H_G(f)$ approximately satisfies a Gibby-Smith criterion (Problem 3.1).

Note that neither waveform is physically realizable; this will be explained later. Before investigating realizable waveform shapers, let us acquaint ourselves with the criteria for jitter-free waveform shaping.

3.1.2 Controlling Jitter

The criterion for jitter-free waveform shaping can be described with the help of functions $\hat{H}(f)$ and $\tilde{H}(f) = H(f)H(f_0 - f)$, as shown in Figure 3.3, where $H(f)$

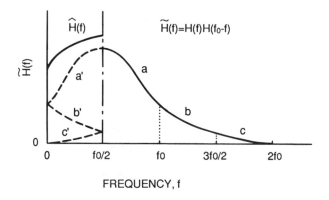

Figure 3.3 Criterion for jitter-free waveform shaping.

represents a spectral density of an isolated pulse (see Section 4.3). The criterion is

$$\arg\{\hat{H}(f)\} = \text{constant}, \quad 0 \le f \le f_0/2 \tag{3.5}$$

where $\hat{H}(f)$ is the waveform function previewed in Section 1.4. That can be defined as

$$\hat{H}(f) = a'(f) + b'(f) + c'(f) \tag{3.6}$$

The amplitude of the waveform function $|\hat{H}(f)|$ does not directly contribute to jitter generation. However, it can enhance pattern dependence of amplitude variation in the recovered clock component, which in turn generates jitter through amplitude to phase conversion.

The amplitude of waveform functions for the raised cosine and Gaussian waveforms mentioned above are compared in Figure 3.4. Note that a much larger enhancement of variation in the amplitude of the clock component is possible with raised cosine shaping as compared with Gaussian shaping.

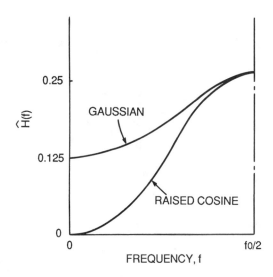

Figure 3.4 Waveform functions for Gaussian and raised cosine shaping.

The phase of waveform function, $\arg[\hat{H}(f)]$, has direct influence on jitter generation. It will be shown in Section 4.3 that waveforms which satisfy (3.5) are broad-sense symmetrical waveforms, defined in Section 1.3. Such waveforms, however, can generate a considerable amount of ISI. Compromised designs will be studied in the next section.

3.1.3 Compromised Design

Let us study, as a compromise, such narrow-sense symmetrical waveform shapers as Gaussian filtering, $H_G(f)$, and raised cosine filtering, $H_R(f)$, as defined by (3.3) and (3.4), respectively. They are not physically realizable because their cut-off slopes increase limitlessly at high frequencies, as seen in Figure 3.5 (note curves marked $\varepsilon_0 = 0$). We can use the following modification to obtain finite cut-off slopes

$$|H_{xm}(f)| = [|H_x(f)| + \varepsilon(f)|H_x(0)|]/[1 + \varepsilon(f)] \tag{3.7}$$

where

$$\varepsilon(f) = \varepsilon_0[1 + (f/f_c)^{2n}]^{1/2} \tag{3.8}$$

and where ε_0 is a sufficiently small number, n is the order-of-cutoff slope, $X = G$ or R, and f_c is usually chosen to be pulse-repetition frequency f_0. Phases of $H_{xm}(f)$

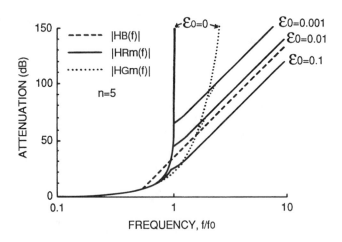

Figure 3.5 Modification of Gaussian and raised cosine filtering. After Y. Takasaki [8].

can be assumed to obey the minimum-phase constraint [4] in practical systems (see Appendix A for the minimum-phase constraint). The fast minimum-phase transfer (FMT) (see Appendix A and [5]) is convenient for calculating phases of $H_{xm}(f)$. Also, let us assume the application of a 50% duty pulse (an RZ pulse) instead of an impulse. Then, parameters like f_g in (3.3) and f_r in (3.4) must be chosen to satisfy the cut-off conditions illustrated in Figure 3.6 and explored in Problem 3.2.

ISI for the RZ pulse response of $H_{Rm}(f)$ and $H_{Gm}(f)$, with a typical cut-off slope ($n = 5$), is shown in Table 3.1. Note that an ε_0 between 0.01 and 0.1 minimizes ISI, although a rather small dependence of ISI on ε_0 is observed for $H_{Gm}(f)$. Some examples of $|H_{Xm}(f)|$ with $n = 5$ are shown for $H_R(f)$ and $H_G(f)$, taking ε_0 as a parameter in Figure 3.5. Note that ε_0 has influence only on out-of-band components of filter responses with rather small amplitudes. Thus, we understand that waveform distortions are mainly caused by delay distortion. A Bessel shaping function $|H_B(f)|$ [6,7] with $n = 5$ is also shown in Figure 3.5 (see Appendix B for $H_B(f)$). Note that, from the viewpoint of minimizing intersymbol interference, $H_B(f)$ can be, substantially, a practical minimum-phase realization for $H_R(f)$ and $H_G(f)$, for a given cut-off slope $n = 5$. This is also true for n values other than 5, as seen in Appendix B.

Maximum ISI, or ISI_M, and maximum jitter J_M for RZ and NRZ pulse responses of $H_B(f)$ are listed in Table 3.2. Maximum jitter J_M is defined in Sections 1.5 and 4.3. Maximum ISI is defined as

$$ISI_M = 2 \sum_{\substack{n=-\infty \\ n \neq 0}}^{\infty} |h_{Bx}(nT_0 + \tau)| / |h_{Bx}(\tau)| \qquad (3.9)$$

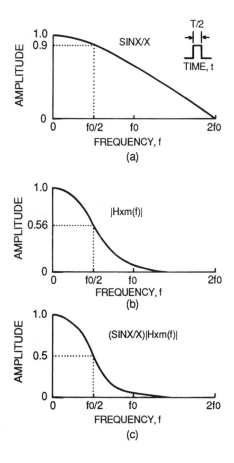

Figure 3.6 Cut-off conditions for an RZ pulse stream; After Y. Takasaki [8]: (a) spectral densities of an RZ pulse; (b) specifications for a shaping filter; and (c) spectral densities of shaped waveform.

where $h_{Bx}(t)$ is RZ or NRZ pulse response of $H_B(f)$, and τ is the time when the pulse peak is encountered. Note that making J_M small requires sufficiently large n with RZ responses. Note also that ISI_M can not be made smaller than $\pm 6\%$. Some improvement can be attained with a predistortion technique, which we will study in Section 5.3.

On the other hand, we can enjoy rather small ISI and jitter by adopting the NRZ pulse. However, we will see in Section 5.2 that the system becomes more susceptible to crosstalk, and the amplitude variation of the clock component can be enhanced, when we adopt the NRZ format.

Table 3.1
Dependence of Intersymbol Interference on ε_0 in Modified Functions

ε_0	Intersymbol Interference(%)	
	$H_{Rm}(f)$	$H_{Gm}(f)$
0.2	3.6	
0.1	2.7	5.5
0.02	2.9	5.0
0.01	4.8	5.0
0.0001		5.3

Table 3.2
Jitter and ISI in Bessel Filtering

Order of Filter n	R Z			NRZ		
	Delay (T)	ISI_M (%)	J_M (Deg.)	Delay (T)	ISI_M (%)	J_M (Deg.)
1	0	>70	26.9	0	9.4	0.1
2	0	10.0	46.8	0	1.6	1.5
3	0.06	6.9	54.5	0.05	1.5	-0.4
4	0.17	5.9	54.0	0.11	1.8	-2.0
5	0.28	6.5	48.6	0.17	2.0	-2.3
6	0.39	6.7	40.6	0.23	2.3	-1.4
7	0.48	6.7	31.8	0.30	2.6	-0.4
8	0.59	6.8	23.4	0.36	2.8	0.2
9	0.66	7.0	16.1	0.41	2.9	0.4
10	0.73	7.5	10.4	0.45	3.1	0.3
11	0.80	8.0	6.2	0.50	3.4	0.2

3.2 EQUALIZER DESIGN

As we discussed in Chapter 1, the use of metallic cables in short distance transmission is expected to become more important. It is well known that metallic cables exhibit square root f characteristics $C(j\omega)$, defined by

$$C(j\omega) = A \exp\{-B\sqrt{j\omega}\} \tag{3.10}$$

That is, when expressed in decibels, amplitude responses of the cable are proportional to the square root of frequency f (Problem 3.3). The variations of metallic cable responses due to ambient temperature and repeater spacing are illustrated in Figure 3.7. The most important frequency in designing equalizers for digital transmission is empirically known to be $f_0/2$, where f_0 is the pulse repetition frequency. It is referred to as equivalent transmission frequency f_e. The standard loss L_0 at f_e is defined in Figure 3.7. Loss variations, $\pm\Delta L$, due to ambient temperatures, are also included in the figure.

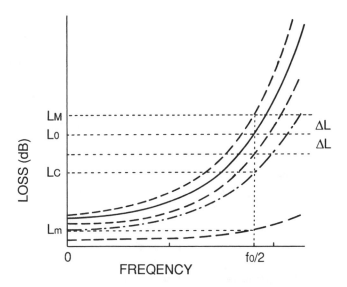

Figure 3.7 Variations of metallic cable responses due to ambient temperature and repeater spacing.

Automatic line build-out (ALBO) is used to adapt the cable loss variation that ranges from minimum, L_m, to maximum, $L_M(=L_0 + \Delta L)$, and is centered at L_c. An ALBO is comprised of a fixed equalizer with a gain of L_c at f_e and a variable equalizer with a variable range of $\pm L_v (L_v = (L_M - L_m)/2)$.

Typical values of L_M and L_V are shown in Table 3.3 with other major parameters. A smaller L_M value is used with twisted-pair cable systems than with coaxial cable systems, because impairments by crosstalks are dominant with the former, and thermal noise determines transmission quality with the latter. On the other hand, a smaller L_V is employed in the coaxial cable system: it is difficult to realize a wide variable range at higher transmission rates (see Section 3.2.2).

Table 3.3
Examples of ALBO Design Parameters

TRANSMISSION MEDIUM	LM (dB)	Lv (dB)	DOMINANT NOISE	TRANSMISSION RATE	REPEATER SPACING
TWISTED PAIR CABLE	40	15	NEAR END CROSSTALK	1.5 Mb/s	2 km (0.65 mm TWISTED PAIR)
COAXIAL CABLE	56	10	THERMAL NOISE	400 Mb/s	1.6 km (2.6/9.5 STANDARD COAX)

The design of amplitude equalizers for digital transmission is rather simple. The use of phase (delay) equalizers is unnecessary in many cases. Therefore, we mainly consider the design of variable equalizers below.

3.2.1 Amplitude and Phase (Delay) Equalizers

The use of amplitude equalizers with minimum-phase transfer functions is acceptable, since the response of metallic cables can be characterized by minimum-phase response (3.10) plus constant delay. Some types of optical fibers also obey the minimum-phase constraint. We can use FMT (see Appendix A) for checking whether a transmission medium incorporates a minimum-phase transfer function.

The use of phase (delay) equalizers can be useful for controlling both jitter and ISI. How to simplify their implementation is the main problem in their design. One useful solution is described in Appendix C [10].

3.2.2 Variable Equalizers

A simple variable equalizer has been proposed by Tarbox [11]. Reduction of equalization errors has also been reported [12]. Bode-type variable equalizers permit us a more theoretical approach [13]. We study the inductorless realization of Bode-type equalizers, and more generalized versions [14] as well, in this section.

Bilinear Variable Equalizer

Variable transfer function $v(f,x)$ simulates a varying transmission characteristic, $y_0(f)^u$, as

$$y_0(f)^u \simeq v(f,x) \tag{3.11}$$

$$u = g(x) \tag{3.12}$$

where the assumption $-1 \leq u \leq 1$ may be practical. Bode-type variable equalizers utilize a bilinear variable transfer function

$$v = (x + y_0)/(xy_0 + 1) \tag{3.13}$$

$$u = (1 - x)/(1 + x), \quad 0 \leq x \leq \infty \tag{3.14}$$

Examples of frequency responses for (3.13) are shown in Figure 3.8. Solid and dotted lines indicate desired characteristics, $y_0(f)^u$, and realizable ones, $v(f,x)$ in (3.11), respectively. Simulation errors can be defined as

$$e(f,u) = 20 \log|y_0(f)^u/v(f,x)| \tag{3.15}$$

Note that simulation errors $e(f,u)$ in (3.15) are zero for three values of u, that is; $u = -1$, 0 and 1 ($x = \infty$, 1 and 0). Maximum errors are encountered half-way between $u = -1$ and 0 ($x = \infty$ and 1), and between $u = 0$ and 1 ($x = 1$ and 0). Errors at $f_e(=f_0/2)$ are plotted in Figure 3.9 by a dotted line. We can understand that using parameter u is convenient for evaluating errors. Usually a maximum allowable error of approximately 1 dB is used to limit degradation in ISI to 10%. In Figure 3.9, y_0 is chosen to be 0.1, that is, the variable range is ±20dB. This variable range selection resulted in maximum errors of around 2.5 dB. Errors increase exponentially with variable range expansion.

Before investigating the reduction of simulation errors, let us study physical realization of the variable transfer function (3.13). It is well known that approximation of y_0 in (3.13) in terms of Bode-type passive networks is extremely complicated. It is also necessary to use inductors in the realization. Fortunately, several

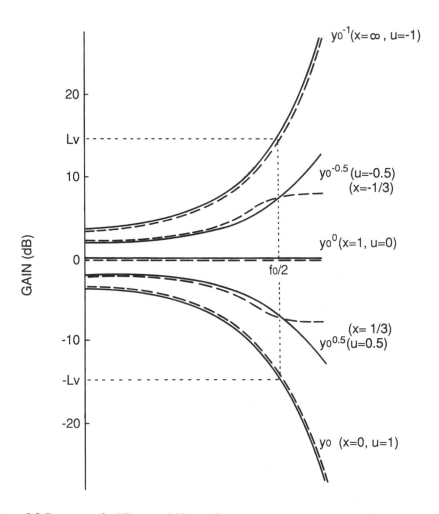

Figure 3.8 Responses of a bilinear variable equalizer.

methods have been developed to make design processes much simpler. One of the most straightforward realizations [15] is shown in Figure 3.10. This type of circuit also suits realizations in terms of integrated circuits, since it does not require the use of inductors. Feedback and feed-forward networks of the same structure can be used. The design of the network is straightforward, since its transfer function is the same as the loss characteristics of the cable. Note that it is difficult to attain the full variable range of x, that is, $0 \leq x \leq \infty$. A physically realizable range may be $(1/x_M) \leq x \leq x_M$, where x_M is a maximum variable range of x. Then the variable range of parameter u in (3.14) is limited to $-u_M \leq u \leq u_M(u_M = (x_M - 1)/(x_M + 1))$ instead of $-1 \leq u \leq 1$.

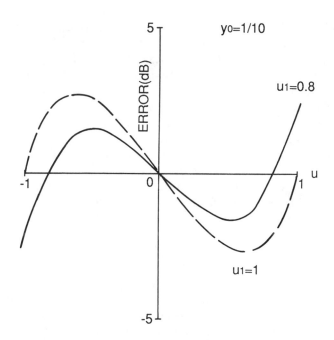

Figure 3.9 Errors caused by bilinear variable equalizers. After Y. Takasaki [14].

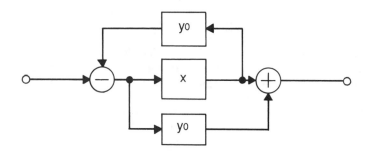

Figure 3.10 Realization of variable equalizer by using feedback and feed-forward. After Y. Takasaki [15].

Example 3.1

A variable range of $0.1 \leq x \leq 10$ can be realized by a combination of an amplifier with a 20 dB gain and an attenuator with a variable range of 0 to 40 dB. It is seen from (3.14) that $0.1 \leq x \leq 10$ makes the parameter u change in a range of $0.8 \gtrsim u \gtrsim -0.8$. This means an 80% utilization of the full variable range is attained. Let us assume $y_0(f) = 0.1$ at $f = f_e$ in (3.13). Then half-variable range L_v in Figure 3.8 is 20 dB. A variable range of ± 20 dB can be attained under the ideal condition of $0 \leq x \leq \infty$, that is, $1 \geq u \geq -1$. A practical condition of $0.1 \leq x \leq 10$, that is, $0.8 \geq u \geq -0.8$, results in a narrower usable variable range of $\pm 0.8 L_v = \pm 16$ dB (Problems 3.4 and 3.5).

It should be noted that the out-band cut-off characteristics of the amplifier in the x element play an important role in amplitude deviations from desired characteristics at higher frequencies. "Out-band" is defined here as a frequency band higher than pulse repetition frequency f_0. It is well known that out-band cut-off shape affects in-band phase distortions. Such phase distortions can, in turn, cause amplitude errors. Equalization errors can be made sufficiently small by using a bandwidth ten times as large as pulse repetition frequency f_0. Simple out-band shaping technology developed in [15] is useful for alleviating such an excessive bandwidth requirement.

Modified Bilinear Equalizer

Now let us study reduction of simulation errors inherent in the bilinear variable transfer function (3.13). Generalized expressions for variable transfer functions are in Appendix D. A generalized bilinear variable transfer function can be given by

$$v = (z_1 x + z_0)/(y_1 x + 1) \tag{3.16}$$

$$u = [(\hat{z}_1 - \hat{y}_1)x + \hat{z}_0]/(\bar{y}_1 x + 1) \tag{3.17}$$

It can be shown that simulation error $e(f,u)$ depends only on the selections of u ($i = 1$, 2 and 3) that force $e(f,u)$ to be zero at each u_i, but that the selections of x ($i = 1$, 2 and 3) do not affect the value of $e(f,u)$ as explored in Problem 3.6.

Example 3.2

We can show that by forcing $e(f,u)$ to be zero at $u = -1$, 0 and 1, and by selecting corresponding x values as $x = \infty$, 1 and 0, respectively, we can get the conventional variable transfer function (3.13). That is, by substituting (u,x) pairs, $(-1,\infty)$, $(0,1)$ and $(1,0)$, in (3.17) we obtain $\hat{z}_0 = 1$, $\hat{y}_1 - \hat{z}_1 = \bar{y}_1 = 1$. Applying these relations in (3.17) results in (3.14). Similarly, by applying (v,x) pairs, (y_0^{-1},∞), $(y_0^0,1)$ and $(y_0,0)$ in (3.16) we obtain $z_0 = y_0$, $z_1 = 1$ and $y_1 = y_0$. Thus we can get (3.13) from (3.16).

Now, notice that zero-forcing of $e(f,u)$ at (u,x) pairs of (u_1,x_1), $(0,1)$ and $(-u_1, x_1^{-1})$ can contribute to the reduction of simulation errors ($u_1 \neq 1$). Applying the above values of (u,x) pairs in (3.16) and (3.17) results in

$$v = (x + \langle y \rangle)/(\langle y \rangle x + 1), \tag{3.18}$$

$$u = u_1(1 + x_1/1 - x_1)(1 - x)/(1 + x), \tag{3.19}$$

$$\langle y \rangle = (y_0^{u_1} - x_1)/(1 - y_0^{u_1}x_1). \tag{3.20}$$

Errors for function (3.18) with $u_1 = 0.8$ and $y_0 = 0.1$ are plotted by a solid line in Figure 3.9. Curves in the figure suggest that function (3.18) is preferred when the extreme values of $x = 0$ and/or $x = \infty$ can not be realized, as is the case in practical applications. A circuit for realizing function (3.18) is shown in Figure 3.11.

Biquadratic Variable Equalizer

The biquadratic variable transfer function (BQ-VTF) can be defined as

$$v = \frac{z_2 x^2 + z_1 x + z_0}{y_2 x^2 + y_1 x + 1} \tag{3.21}$$

$$u = \frac{(\hat{z}_2 - \hat{y}_2)x^2 + (\hat{z}_1 - \hat{y}_1)x + \hat{z}_0}{\tilde{y}_2 x^2 + \tilde{y}_1 x + 1} \tag{3.22}$$

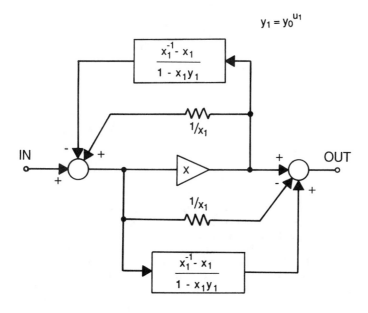

Figure 3.11 Realization of modified bilinear variable equalizer. After Y. Takasaki [14].

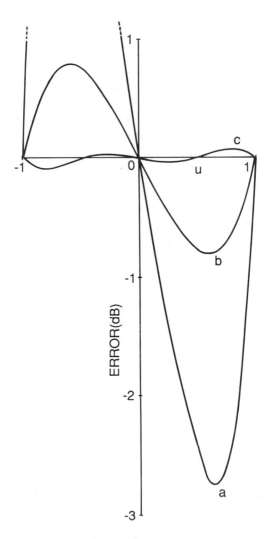

Figure 3.12 Capacity of biquadratic variable equalizer for error reduction: (a) bilinear variable equalize (b) two cascaded bilinear variable equalizers; and (c) biquadratic variable equalizer. *Sourc* Takasaki Y., *et al.*, "High-Precision Inductorless Variable Equalizers with Wide Variab Range," *IEEE Trans. Circuits and Syst.*, Vol. CAS-24, No. 12, December 1977, pp. 704 708. Reprinted with permission. Copyright © 1977 IEEE.

Analyses of BQ-VTF are rather complicated. First, it is advisable to incorporate the following practical simplifications

$$\text{ZFs} \quad \text{at} \quad u = 1, \quad 1/2, \quad 0, \quad -1/2, \quad -1$$

and symmetry with regard to $u = 0$, for example,

$$v(f,x(u)) = v(f,x(-u))^{-1}$$

It can be shown that $e(f,u)$ depends on both u_i and x_i. One typical BQ-VTF is described below. In addition to zero-forcings in terms of u parameter defined above, zero-forcings at $x = 0$, x_2, 1, x_2^{-1}, and ∞ are applied, which yields

$$v = \frac{x^2 + y_1 x + y_0}{y_0 x^2 + y_1 x + 1} \tag{3.23}$$

$$u = \frac{-x^2 + 1}{x^2 + \tilde{y}_1 x + 1}, \quad 0 \leq x \leq \infty \tag{3.24}$$

$$y_1 = -x_2(1 + y_0) + (1 - x_2^2)x_2^{-1} y_0^{1/2} \tag{3.25}$$

$$\tilde{y}_1 = (1 - 3x_2^2)/x_2 \tag{3.26}$$

Drastic error reduction can be attained by using (3.23). Error curves for different types of variable transfer functions are compared in Figure 3.12 for $L_v = 20$ dB ($y_0 = 0.1$). Curve (a) shows errors caused by the conventional bilinear variable function (3.13). Errors can be reduced by cascading two bilinear equalizers, curve (b), and taking advantage of the fact that errors decrease exponentially as L_v decreases. We can obtain very small simulation errors, curve (c), by choosing $x_2 = 1/3$ in (3.25) and (3.26). A physical realization of BQ-VTF (3.23) is shown in Figure 3.13. Further details of this type of variable equalizer design can be found in [16].

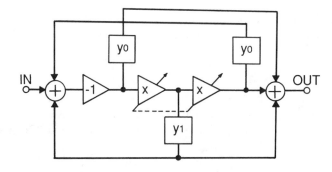

Figure 3.13 Realization of biquadratic variable equalizer. After Y. Takasaki [16].

Linear Variable Equalizer

Variable transfer functions described so far require a feedback path for their physical realization. The use of feedback may not be desirable with very high-speed applications. Linear and quadratic variable transfer functions (L-VTF and Q-VTF) are useful in alleviating such a problem. In Appendix D, the linear (first-order) variable transfer function can be defined as

$$v = z_1 x + z_0 \qquad (3.27)$$

$$u = \hat{z}_1 x + \hat{z}_0 \qquad (3.28)$$

The zero-forcing of simulation error $e(f,u)$ expressed by (3.15) at $(x,u) = (x_1, u_1)$ and (x_2, u_2) yields

$$z_1 = (y_0^{u_1} - y_0^{u_2})/(x_1 - x_2) \qquad (3.29)$$

$$z_0 = (x_1 y_0^{u_2} - x_2 y_0^{u_1})/(x_1 - x_2) \qquad (3.30)$$

$$\hat{z}_1 = (u_1 - u_2)/(x_1 - x_2) \qquad (3.31)$$

$$\hat{z}_0 = (x_1 u_2 - x_2 u_1)/(x_1 - x_2) \qquad (3.32)$$

Substituting these co-efficients in functions (3.27) and (3.28), and expressing v in terms of u yields

$$v = \frac{y_0^{u_1} - y_0^{u_2}}{u_1 - u_2} u + \frac{u_1 y_0^{u_2} - u_2 y_0^{u_1}}{u_1 - u_2}, \qquad (3.33)$$

As seen from (3.33), $e(f,u)$ does not depend on the selection of x_1 and x_2.

Zero-forcing of errors for function (3.27) at $(x,u) = (0,1)$ and $(1,-1)$ yields

$$v = (y_0^{-1} - y_0)x + y_0 \qquad (3.34)$$

Simulation errors $e(f,u)$ are shown in Figure 3.14 by a dotted line for $y_0 = 1/2$. A physical realization of this type of variable transfer function is shown in Figure 3.15. Now, let us study zero-forcing of errors at $(x_1, 0.75)$ and $(x_2, -0.75)$, which yields:

$$v = \frac{y_0^{0.75} - y_0^{-0.75}}{x_1 - x_2} x + \frac{x_1 y_0^{-0.75} - x_2 y_0^{0.75}}{x_1 - x_2} \qquad (3.35)$$

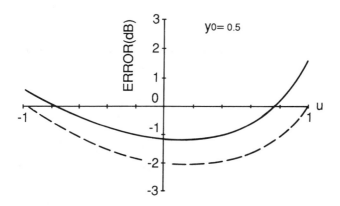

Figure 3.14 Errors caused by linear variable equalizers. After Y. Takasaki [14].

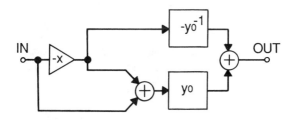

Figure 3.15 Realization of linear variable equalizer.

Error curves for (3.35) are also plotted in Figure 3.14 by a solid line. It is evident that errors in decibel are halved when compared with function (3.34). A realization for function (3.35) is shown in Figure 3.16.

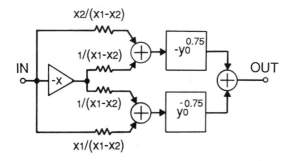

Figure 3.16 Realization of modified linear variable equalizer. After Y. Takasaki [14].

Quadratic Variable Equalizer

The quadratic variable transfer function (Q-VTF) can be expressed as

$$v = z_2 x^2 + z_1 x + z_0 \tag{3.36}$$

$$u = \hat{z}_2 x^2 + \hat{z}_1 x + \hat{z}_0 \tag{3.37}$$

It can be shown that zero-forcings at (x_1, u_1), (x_2, u_2) and (x_3, u_3) lead to an $e(f, u)$ that depends not only on u_i but also on $x_i (i = 1, 2, 3)$. The conventional Q-VTF [17] can be expressed as

$$v = (2y_0 - 4 + 2y_0^{-1})x^2 - (3y_0 - 4 + y_0^{-1})x + y_0 \tag{3.38}$$

$$u = -2x + 1 \tag{3.39}$$

These functions are obtained as results of zero-forcings for function (3.36) at $(x, u) = (0, 1)$, $(1/2, 0)$ and $(1, -1)$. Errors are plotted in Figure 3.17 by a solid line for $y_0 = 1/4$.

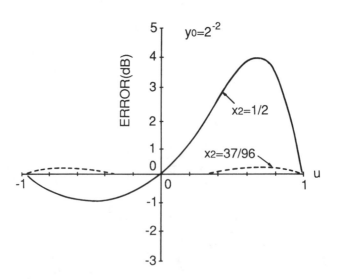

Figure 3.17 Errors caused by quadratic variable equalizers. After Y. Takasaki [14].

Next, zero-forcings at $(0,1)$, $(x_2,0)$ and $(1,-1)$ are considered to evaluate the effect of x_2 on error reduction, which yields

$$v = \frac{1 - y_0}{x_2 - 1}\left[\left(x_2^{-1} - \frac{1 + y_0}{y_0}\right)x^2 - \left(x_2^{-1} - \frac{1 + y_0}{y_0}x_2\right)x\right] + y_0 \quad (3.40)$$

$$u = \frac{1}{1 - x_2}(x_2^{-1} - 2)x^2 + \frac{1}{1 - x_2}(-x_2^{-1} + 2x_2)x + 1 \quad (3.41)$$

Drastic error reduction can be attained by choosing $x_2 = 37/96$, shown in Figure 3.17 by a dotted line. An example for realizing this type of Q-VTF is shown in Figure 3.18.

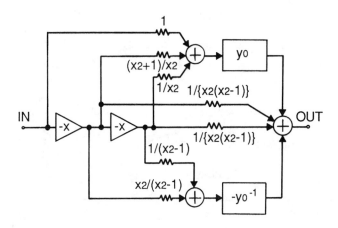

Figure 3.18 Realization of modified quadratic variable equalizer. After Y. Takasaki [14].

3.3 INFLUENCE OF SHAPING ERRORS

Shaping errors due to approximations in equalizer design as well as deviations of component values can result in increased ISI and jitter. Theory developed by Wheeler [18] is convenient for evaluating the influence of such errors on degradations.

Figure 3.19 outlines the theory; section (a) shows that a pair of echoes of the same polarity causes cosine-shaped amplitude errors in the frequency domain. No phase errors are encountered in this case. On the other hand, a pair of echoes with opposite polarities, shown in (b), causes only phase errors, provided ε is small enough (Problem 3.7).

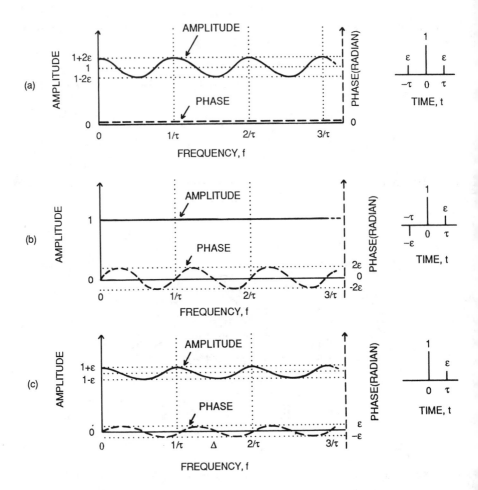

Figure 3.19 Evaluation of shaping errors in terms of echoes: (a) shaping errors due to a pair of echoes of the same polarity; (b) shaping errors due to a pair of echoes of opposite polarities; and (c) shaping errors due to a single echo.

In most practical systems, where the minimum-phase law holds true, the main response is never preceded by an echo as shown in (c); that case can be considered a linear sum of (a) and (b). Example 3.3 below shows how we can apply this theory to evaluate the amount of ISI and jitter caused by shaping errors.

Example 3.3

Let us assume $\varepsilon = 0.05$ and $\tau = 1.5T$ in Figure 3.19(c). Then, pulse repetition frequency $f_0 (= 1/T)$ can be indicated by Δ in that figure. We can see that the am plitude and phase deviations of ± 0.45 dB and $\pm 3°$, respectively, are encountered in the frequency domain. Note that a worst-case ISI of $\pm 10\%$ is caused by an echo with $\varepsilon = 0.05$ (Problem 3.8). Note also that a worst-case jitter of $\pm 6°$ can be generated by this amount of shaping error (Problem 5.1).

More complicated shaping errors can be analyzed by expanding the theory with the incorporation of multiple echoes instead of the single echo in Figure 3.19(c).

PROBLEMS

Problem 3.1 Show that the Gaussian filtering in Figure 3.2 is a good approximation to a Gibby-Smith criterion.

Problem 3.2 A filter is designed to satisfy Nyquist's criterion at $f_0/2$ for the response of an NRZ pulse. Show that the frequency response of the filter is -2 dB at $f_0/2$ (see Figure 3.6).

Problem 3.3 Show that (3.10) exhibits a square root f response.

Problem 3.4 Show examples of $y_0(f_e)$ and an x element that can attain $L_v = 20$ dB.

Problem 3.5 Show an example of an x element that can be electronically controlled.

Problem 3.6 Show that $e(f,u)$ for (3.16) does not depend on the selections of x_i.

Problem 3.7 Provide a mathematical explanation for Figure 3.19(b).

Problem 3.8 Show that a worst-case ISI of 10% is caused by an echo with $\varepsilon = 0.05$.

REFERENCES

1. Nyquist H., Certain Topics in Telegraph Transmission Theory, *AIEE Trans. (Commun. Electron)*, Vol. 47, April 1928, pp. 617–644.
2. Gibby R.A., and J.W. Smith, Some Extensions of Nyquist's Telegraph Transmission Theory, *Bell Syst. Tech. J.*, Vol. 44, September 1965, pp. 1487–1510.
3. Sunde E.D. Theoretical Fundamentals of Pulse Transmission, *Bell Syst. Tech. J.*, Vol. 33, May–July 1954.
4. Bode H.W., Network Analysis and Feedback Amplifier Design, Princeton, N.J.: D. Van Nostrand Co., Inc., 1945.
5. Aoki K., unpublished memorandum, November 1970.
6. Thompson W.E., Network with Maximally Flat Delay, *Wireless Eng.*, Vol. 29, 1952.
7. Storch L., Synthesis of Constant-Time Delay Ladder Networks Using Bessel Polynomials, *Proc. IRE,* Vol. 42, November 1954, pp. 1666–1675.
8. Takasaki Y., Optimizing Pulse Shaping for Baseband Digital Transmission with Self-Bit Synchronization, *IEEE Trans. Commun.*, Vol. COM-28, No. 8, August 1980, pp. 1164–1172.
9. Mayo J.S., A Bipolar Repeater for Pulse Code Modulation Signals, *Bell Syst. Tech. J.*, Vol. 41, January 1962.
10. Takasaki Y., and N. Maeda, Baseband Pulse Transmission in a Chain of Analog Repeaters, *IEEE Trans. Commun.*, Vol COM-22, November 1974, pp. 1823–1835.

11. Tarbox R.A., A Regenerative Repeater Utilizing Hybrid Integrated Circuit Technology, *Int'l. Conf. on Commun.* 1969, Paper #69CP403-COM.

12. Anuff A., *et al.*, A New 3.152 Mb/s Digital Repeater, Proc. Int'l. *Conf. on Commun.* 1975, pp. 39-10–39-13.

13. Bode H.W., Variable Equalizer, *Bell Syst. Tech. J.*, Vol. 17, No. 2, 1938, pp. 229–244.

14. Takasaki Y., Generalized Theory of Variable Equalizers, *Proc. 1979 ISCAS*, pp. 146–149.

15. Takasaki Y., *et al.*, Inductorless Variable Equalizers Using Feedback and Feedforward, *IEEE Trans. Circuits and Syst.*, Vol. CAS-23, No. 6, June 1976, pp. 389–394.

16. Takasaki Y., *et al.*, High-Precision Inductorless Variable Equalizers with Wide Variable Range, *IEEE Trans. Circuits and Syst.*, Vol. CAS-24, No. 12, December 1977, pp. 704–708.

17. Chen W.I.H., Analysis of Adjustable Multipath Network for Use in Building Out Transmission-Line Skin-Effect Losses, *IEEE Trans. Circuit Theory*, Vol. CT-18, No. 4, July 1971, pp. 436–443.

18. Wheeler H.A., The Interpretation of Amplitude and Phase Distortion in Terms of Paired Echoes, *Proc. IRE*, Vol. 27, June 1939, pp. 359–385.

Chapter 4
Clock Recovery

We study principles of clock transmission to understand the applicability of self-timing based on nonlinear extraction. Then, we give an overview of the design of clock recovery systems, utilizing such narrow-band filters as tank circuits, phase-locked loops (PLL), and surface acoustic wave (SAW) filters.

Next, we study the theory of nonlinear clock recovery to understand the usefulness of square-law nonlinearity for analyzing the influence of pulse patterns and waveforms on the behavior of the recovered clock. The concept of waveform function and pattern function is investigated to show that the amplitude and phase of a recovered clock component can be expressed by an inner product of these functions. This allows us to analyze the influence of waveforms on jitter behavior independently from the influence of pulse patterns.

Finally, we study simple examples to understand the efficiency of jitter analyses in terms of waveform and pattern functions.

4.1 PRINCIPLES OF CLOCK TRANSMISSION

Clock transmission schemes are listed in Table 4.1. Although simple timing circuits can be used with external timing schemes, adjusting clock phase is a major problem. Clock signal can either be superposed at the spectral null of an information signal, or transmitted via a separate medium. This type of scheme may be worthy of consideration in short distance transmissions like intra-office and interconnect systems, where the delay difference does not cause significant problems.

Most practical systems assume a self-timing approach. Linear clock extraction is possible if a discrete clock component is available at clock frequency f_0. Let us assume a random pulse sequence as shown in Figure 4.1(a). This pulse stream can be divided into two components; periodic component (b), and nonperiodic component (c). Spectra of these components are indicated by B and C, respectively, in the

Table 4.1
Clock Transmission Schemes

External Timing

    Superposed Transmission

    Separate Transmission

Self Timing

    Linear Extraction

    Nonlinear Extraction

frequency domains shown to the figures' right. It is seen that we must use the Gibby-Smith criterion (curve G) (See Section 3.1.1) in design instead of Nyquist's (curve N) to make the clock component available at f_0. In such a case, however, the quality of the clock component can be susceptible to high-frequency noise. Also time cross-hair budget can be reduced (Problem 4.1).

These problems can be avoided by utilizing nonlinear extraction schemes. The differentiation of a pulse stream usually precedes nonlinear processing, as shown in Figure 4.1(a) to (g). Pulse stream (a) is delayed by one timeslot (waveform (d)) and subtracted from the original stream (a) to result in differentiated stream (e). This is equivalent to the $\sin x$ filtering shown in the figure by a dotted line in the frequency domain to the right. A Nyquist filtering (N) is then applied to this stream to produce stream (f). Note that this is similar to the modified duobinary class 2 we studied in Section 2.5. Composite filtering (F) of the $\sin x \cos^2 x$ type, can be used to this end.

The full-wave rectification of waveform (f) results in waveform (g). A new, continuous spectrum is thereby obtained, and a discrete line spectrum absent from the original wave is created. It is apparent that mid-band components around $f_0/2$ in spectra F are converted into the discrete line spectrum at f_0.

One problem with this scheme is that a considerable amount of timing jitter can be generated if a shaped waveform is not sufficiently symmetrical. This is explained in Figure 4.2, where waveform (b) is created by a symmetrical waveform (a), followed by a pulse with the same waveform of opposite polarity. The full-wave rectification of waveform (b) results in waveform (c), which is comprised of sym-

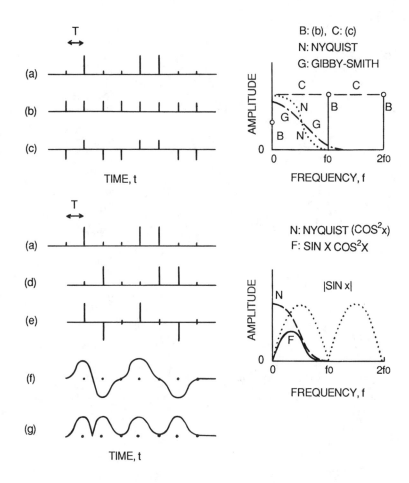

Figure 4.1 Principle of clock recovery: (a) random pulse sequence; (b) discrete spectral component (periodic component); (c) continuous spectral component (non-periodic component); (d) stream (a) delayed by one timeslot; (e) stream (a) minus stream (d) (differentiated waveform); (f) response of stream (e) to a raised cosine filter; and (g) full-wave rectified version of stream (f).

metrical waveforms as indicated by dotted and dashed lines. Therefore, the phases of the clock components extracted from pulse (a) and composite pulse (b) are shown to be the same.

On the other hand, when an asymmetrical waveform like that shown in Figure 4.2(a′) is used, an extra-asymmetrical waveform, shown by the dashed curve in (c′), is generated. Consequently, the phase of the clock component can vary depending on pulse patterns. Analyses of and solutions to this problem are provided in Chapter 5.

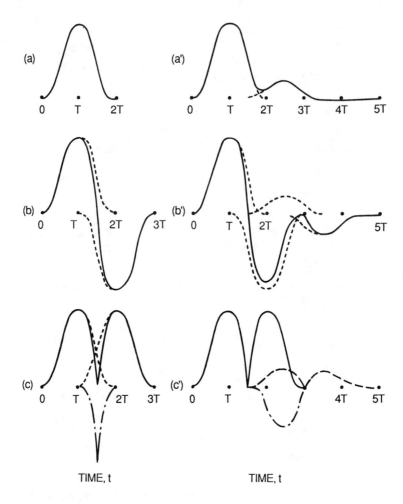

Figure 4.2 Jitter generation due to asymmetrical pulse shape: (a, a') symmetrical and asymmetrical pulse waveforms, respectively; (b, b') pulse followed by another pulse with opposite polarity; and (c, c') full-wave rectified waveforms with in-phase and out-of-phase components.

4.2 DESIGN OF CLOCK RECOVERY SYSTEMS

A full-wave rectified pulse stream is applied to a clipping circuit before extracting the clock component. The optimum clipping ratio is studied in Figure 4.3. It is shown that clipping ratio ρ_c in the neighborhood of 0.3 maximizes the extracted component. Note that the dynamic range of the extracted clock component can be compressed by the ac coupling of the rectified waveform before clipping (Figure 4.3(b)).

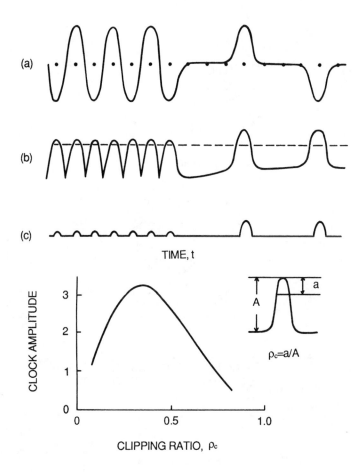

Figure 4.3 Optimization of pulse clipping for the extraction of clock component: (a) pulse stream with AMI format; (b) response of full-wave rectified stream to ac coupling; and (c) stream (b) after clipping.

Several types of narrow-band filters for clock component extraction are listed in Figure 4.4. One of the typical filters is a tank circuit shown in (a), and explored in Problem 4.2. This circuit features simple implementation [1]. It is useful for lower-speed applications. PLLs(b) (also see Problem 4.3) apply to systems with low to medium clock rates [2,3]. For higher-speed applications, SAW filters [4] or dielectric resonators (DR) can be used (c). The single-tuned filtering in (a) will be used in the following analyses. Almost the same results can be obtained with other filtering schemes, provided they are designed to satisfy specifications and requirements suitable for practical applications.

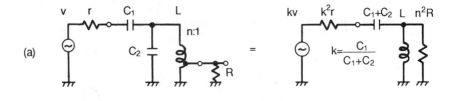

(a)

VCO: VOLTAGE CONTROLLED OSCILLATOR

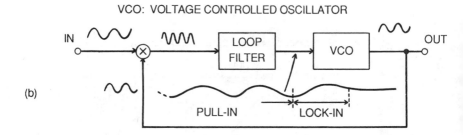

(b)

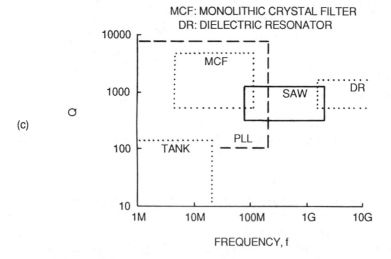

(c)

Figure 4.4 Narrowband filters for clock extraction: (a) tank circuit; (b) phase lock loop (PLL); and (c) applicability of narrow-band filters for the parameters of clock extraction.

4.3 NONLINEAR CLOCK RECOVERY

The proper choices of nonlinearity and waveform are necessary to minimize jitter in the presence of pulse overlaps. Let us study timing extraction through a general nonlinearity

$$y(x) = \sum_{n=0}^{N} r_n x^n \qquad (4.1)$$

and in terms of pulse burst

$$x(t) = \sum_{i=1}^{M} a_i h(t - iT) \qquad (4.2)$$

where $h(t)$, a_i, T, and MT are pulse shape, pulse amplitude, pulse repetition period, and pattern period, respectively. In [5], conditions necessary for jitter-free timing extraction are discussed. If a_i in (4.2) takes continuous amplitude as in a pulse amplitude modulation (PAM) system, it is necessary for $h(t)$ to be symmetrical and

$$y(x) = (r_0 + r_1 x) + r_2 x^2 \qquad (4.3)$$

where r_0, r_1 and r_2 are arbitrary constants. It is also sufficient as shown in [6]. It should be self-evident in Figure 4.5 that simple full-wave retification can cause jitter in multilevel systems even if symmetrical waveform shaping is applied (Problem 4.4). If a_1 takes binary $(a, -a)$ or ternary $(a, 0, -a)$ amplitudes, as in typical pulse code modulation (PCM) systems, it is necessary that

$$y(x) = (r_0 + r_1 x) + \sum_{l=1}^{L} r_{2l} x^{2l} \qquad (4.4)$$

for symmetrical waveform $h(t)$.

The nonlinearity (4.3) is indispensable in PAM, correlative [7, 8], or multilevel [9] transmission systems. This type of nonlinearity is convenient even with binary or ternary systems, since it can greatly simplify the analysis of clock recovery without sacrificing the generality of design theory, as will be discussed later. Therefore, we mainly assume square-law analysis in this book.

Timing Extraction by Square-Law Device

Let us define an impulse pattern as

$$w(t) = \sum_{i=0}^{M-1} a_i \delta(t - iT) \qquad (4.5)$$

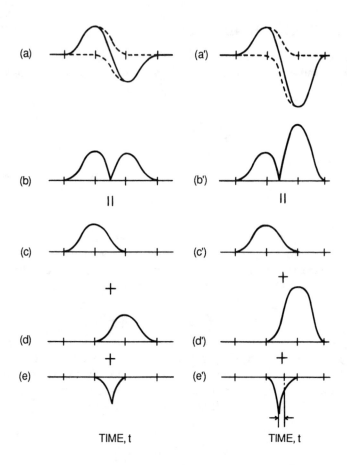

Figure 4.5 Jitter generation due to full-wave rectification of multilevel pulse stream that comprises symmetrical pulses: (a, a′) pulse followed by another pulse with opposite polarity; (b, b′) full-wave rectified waveform; (c, c′) the first components of waveforms in (b) and (b′); (d, d′) the second components of waveforms (b) and (b′); and (e, e′) the third components of waveforms (b) and (b′).

where $\delta(t)$ is a delta function. It is convenient for analysis to describe a pulse train $x(t)$ by a convolution of an impulse pattern $w(t)$ and a pulse shape $h(t)$ as $x(t) = w(t) \otimes h(t)$, which in the frequency domain becomes a product $X(f) = W(f) \cdot H(f)$, where

$$W(f) = \sum_{i=0}^{M-1} a_i \exp\left(-2\pi j \frac{f}{f_0} i\right) \tag{4.6}$$

and $f_0 = 1/T$ (pulse repetition frequency).

Frequency spectrum $x^{\langle 2\rangle}(f)$ of a squared pulse train $x^2(t)$ is given by

$$x^{\langle 2\rangle}(f) = \int_{-\infty}^{\infty} X(f - f_1)X(f_1)df_1 \qquad (4.7)$$

Consider timing extraction from the Nth harmonic of f_0 in $x^{\langle 2\rangle}(f)$. Making use of the relations

$$W(f_0 - f)W(f) = |W(f)|^2 \qquad (4.8)$$

$$|W(f + nf_0)|^2 = |W(f)|^2, \ n \text{ integer} \qquad (4.9)$$

$$|W(f_0/2 + f)|^2 = |W(f_0/2 - f)|^2 \qquad (4.10)$$

and after some manipulation we obtain

$$x^{\langle 2\rangle}(Nf_0) = 2 \int_{0}^{f_0/2} |W(f)|^2 \hat{H}(Nf_0, f)df \qquad (4.11)$$

where

$$\hat{H}(Nf_0, f) = \sum_{n=-\infty}^{\infty} H(Nf_0 - \overline{f + nf_0})H(f + nf_0) \qquad (4.12)$$

Since the functions $|W(f)|^2$ and $\hat{H}(Nf_0, f)$ in (4.11) depend only on pulse patterns and pulse waveforms, respectively, we will call them "pattern function" and "waveform function" below. Equation (4.11) shows that the magnitude of a timing wave can be described by the inner product of pattern function and waveform function (over a frequency range $0 - f_0/2$). This expression separates the effects of pulse pattern and pulse shapes on the amplitude and the phase of the timing wave, and offers insight to their contribution to timing jitter.

Pattern Function

As seen from (4.6), (4.9), and (4.10), the pattern function $|W(f)|^2$ is a power spectrum of the impulse pattern $w(t)$, has a period f_0, and is even symmetric about $f_0/2$. Therefore, we can easily draw the shape of the function (Figure 4.6). The fact that the pattern function is real, not negative, makes qualitative estimation of the timing wave by (4.11) very simple.

There are some impulse trains $w(t)$ corresponding to an arbitrary pattern function for PAM pulse trains. Since the pattern function contains no phase information,

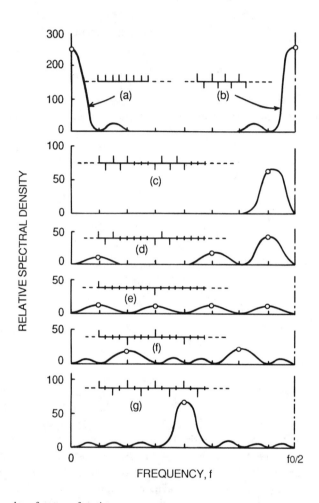

Figure 4.6 Examples of pattern functions.

many impulse patterns can correspond to a single pattern function. Coding schemes restrict the shape of pattern function. Constrained bipolar coding [10], or AMI, for example, emphasize the function in the vicinity of $f_0/2$ when pulse density is high, as shown in Figures 4.6(b) and (c). A sparse pulse train makes the pattern function uniform, as shown in (e). We can use these facts to reduce amplitude variation of the timing wave as shown in Section 5.3.2.

Waveform Function

The waveform function reveals the effects of pulse overlaps on the timing wave. Let us confine our attention to $\hat{H}(f_0, f)$, and denote it $\hat{H}(f)$ below. The relation between the spectrum of a pulse waveform and its waveform function can be interpreted as illustrated in Figure 4.7. As seen from that figure and (4.12), a waveform function

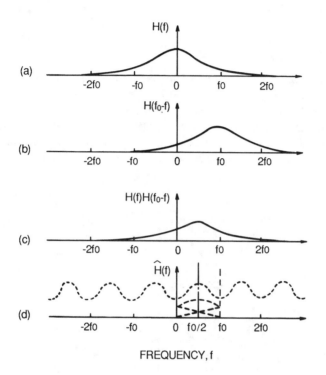

Figure 4.7 Interpretation of waveform function.

is a constant when a function $F(f)$

$$F(f) = \tilde{H}(f + f_0/2)/\tilde{H}(f_0/2) \tag{4.13}$$

$$\tilde{H}(f) = H(f)H(f_0 - f) \tag{4.14}$$

satisfies the Gibby-Smith criterion (Problem 4.5). For instance, $\hat{H}(f)$ is a constant for an ideal low-pass function $H(f) = 1(|f| < f_0)$, $= 0$ (otherwise) or for an exponential low-pass function $H(f) = \exp(-a|f|)(|f| < f_0)$, $= 0$ (otherwise). In these cases, pulse overlaps have no influence on the timing wave. Amplitudes of a waveform function, however, need not be constant. Conversely, we can reduce amplitude variation of the timing wave by properly shaping amplitude and frequency characteristics of a waveform function, as shown in Section 5.3.2. On the other hand, the angles of a waveform function must be constant, since otherwise the phase of the timing wave fluctuates with the variation of digital patterns. The upper bound to jitter is given by the difference between the maximum and minimum angles in a waveform function.

Example 4.1

Simplified examples of spectral expression $H(f)$ for pulse waveforms that can represent some practical cases are shown in Figure 4.8(a). Solid and dotted lines represent 100% and 50% roll-off, respectively. In these cases, waveform function $\hat{H}(f)$ can be given simply by a product $H(f)H(f_0 - f)$. Note in Figure 4.8(c) that the amplitude and phase of roll-off characteristics can have significant influence on a waveform function in the vicinity of zero frequency. A pulse sequence with a

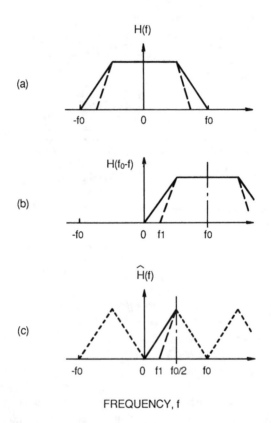

FREQUENCY, f

Figure 4.8 Simple examples of pulse spectra and corresponding waveform functions: (a) spectral expression $H(f)$ of a pulse waveform; (b) $H(f_0 - f)$; and (c) waveform function $\hat{H}(f)$.

attern function that has major components in a lower-frequency band, as in Figure
.6(a), can cause considerable degradation to the quality of the recovered clock.
uch a problem can be alleviated by either employing line coding that can suppress
 succession of marks, or differentiating a pulse stream before applying it to a non-
near device (Figure 4.1(d–e)).

PROBLEMS

Problem 4.1 Show that waveform shaping based on Gibby-Smith criteria can result
in a smaller time crosshair budget than one based on Nyquist shaping.

Problem 4.2 Explain how the tank circuit on the left in Figure 4.4(a) can be expressed by the equivalent circuit shown on the right.

Problem 4.3 Speculate on how the phase-locked-loop (PLL) shown in Figure 4.4(b)
works.

Problem 4.4 Show that full-wave rectification can cause jitter even if symmetrical
waveform shaping is applied as shown in Figure 4.5.

Problem 4.5 Explain why $\hat{H}(f)$ is a constant when $F(f)$ in (4.13) satisfies a Gibby-
Smith criterion.

REFERENCES

1. Mayo J.S., A Bipolar Repeater for Pulse Code Modulation Signals, *Bell Syst. Tech. J.*, Vol. 41,
    Jan. 1962, pp. 25–97.
2. Runge P.K., Phase-Locked Loops with Signal Injection for Increased Pull-In Range and Reduced
    Output Phase Jitter, *IEEE Trans. Commun.*, Vol. COM-24, No. 6, June 1976, pp. 636–644.
3. Bellisio J.A., A New Phase-Locked Timing Recovery Method for Digital Regenerators, *Conf.
    Record ICC '76*, Philadelphia PA, June 1976, pp. 10-17–10-20.
4. Rosenberg R.L., C. Chamzas and D.A. Fishman, Timing Recovery with SAW Transversal Fil-
    ters in the Regenerators of Undersea Long-Haul Fiber Transmission Systems, *IEEE J. Lightwave
    Tech.*, Vol. LT-2, No. 6, December 1984, pp. 917–925.
5. Takasaki Y., Timing Extraction in Baseband Pulse Transmission, *IEEE Trans. Commun.*, Vol.
    COM-20, October 1972, pp. 877–884.
6. Takasaki Y., Systematic Jitter Due to Imperfect Equalization, *IEEE Trans. Commun. Technol.*,
    Vol. COM-19, December 1971, pp. 1,275–1,276.
7. Lender A., Correlative Digital Communication Techniques, *IEEE Trans. Commun. Technol.*,
    Vol. COM-12, December 1964, pp. 128–135.
8. Kretzmer E., Binary Data Communication by Partial Response Transmission, *1st Ann. IEEE
    Communications Conv.*, Boulder, CO, June 1965, pp. 451–455.
9. Pierce J.R., Information Rate of a Coaxial Cable with Various Modulation Systems, *Bell Syst.
    Tech. J.*, Vol. 45, October 1966.
10. Aaron M.R., PCM Transmission in the Exchange Plant, *Bell Syst. Tech. J.*, Vol. 41, January
    1962.

Chapter 5
Jitter Generation

We first study the definition of jitter to distinguish timing jitter from alignment jitter, random jitter from systematic jitter, and type A jitter from type B jitter. We also study jitter sources like waveform distortion, amplitude-to-phase conversion, mistuning, finite Q, and nonzero pulsewidth. We also explore some examples of jitter-line structures.

We then focus on jitter due to waveform distortion. We study waveform functions for Bessel filtering to analyze the influence of approximation errors in waveform shapers. Jitter behaviors with RZ and NRZ pulse streams are compared. We also study the influence of deviations in frequency response (realization errors) to obtain criteria for jitter-free waveform shaping in terms of echo theory.

Finally, we study how to control jitter generation. We show that broad-sense symmetrical waveform shaping is efficient for reducing jitter due to waveform distortion. This type of shaping, however, adversely affects ISI. We study a solution to this problem, a combined predistortion and tail-shaping technique, to attain simultaneous reduction of jitter and ISI.

5.1 DEFINITION OF JITTER

We can classify jitter as listed in Table 5.1. Timing jitter can be defined as phase deviation J_t in Figure 5.1 (*i*) of a clock pulse from a reference clock (0). Alignment jitter J_a is defined as the phase difference between an imaginary clock component, carried by a pulse stream input to a regenerative repeater, and a recovered clock component. Spacing jitter is the deviation of clock-pulse spacing between adjacent pulses. Only timing jitter is considered in this section. Discussions on alignment jitter will be raised in succeeding sections.

Timing jitter can be classified as random and systematic. Systematic timing jitter is pattern-dependent. Therefore, the same jitter is introduced at each regenerator in a chain [1]. Random jitter is noise-dependent.

Table 5.1
Classification of Jitter

Classification I

    (a) Timing Jitter (Absolute Jitter)

    (b) Alignment Jitter

    (c) Spacing Jitter

Classification II

    (a) Random Jitter (Noise Dependent)

    (b) Systematic Jitter (Pattern Dependent)

        (i) Type A

           Jitter due to finite Q

           Jitter due to non-zero pulse width

           Jitter due to mistuning

        (i i) Type B

           Jitter due to waveform distortion

           Amplitude to phase conversion

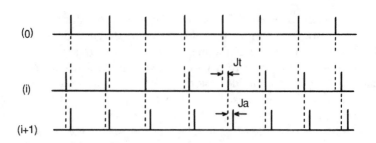

Figure 5.1 Definition of Jitter.

Systematic timing jitter can be classified as type A and type B [2]. While the latter includes low-frequency jitter components, the former does not; we will discuss further details in Section 6.1. The five sources of pattern-dependent jitter are also listed in Table 5.1.

Amplitude-to-phase conversion can be understood by referring to Figure 5.2. In that figure, (a) shows the response of a tank circuit in which Q = 100 to a rectified

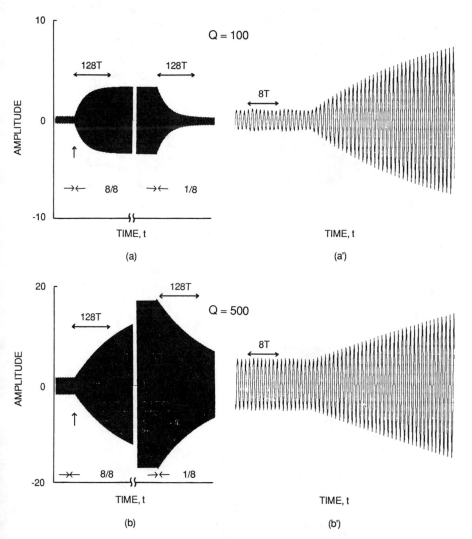

Figure 5.2 Amplitude variations in the extracted clock component due to pattern change (1/8–8/8); (a, a') $Q = 100$, and (b, b') $Q = 500$.

pulse train as in Figure 4.3(c). Pulse pattern changes from 1/8 to 8/8 (all-mark) and then returns to 1/8. The part of the waveform indicated by the arrows is expanded in the time axis and is shown in Figure 5.2(a′). Figures 5.2(b) and (b′) show responses for Q = 500. Note that the amplitude variation in the extracted clock component is so large that any part of the nonlinear circuit following the tuned circuit will perturb the zero-crossings of the timing wave and generate jitter.

Jitter generation due to finite Q, nonzero pulsewidth, mistuning, and waveform distortion is studied in Figure 5.3 (Q = 100) and Figure 5.4 (Q = 500). Figures 5.3(a) and 5.4(a) show that jitter can be generated even in the absence of mistuning

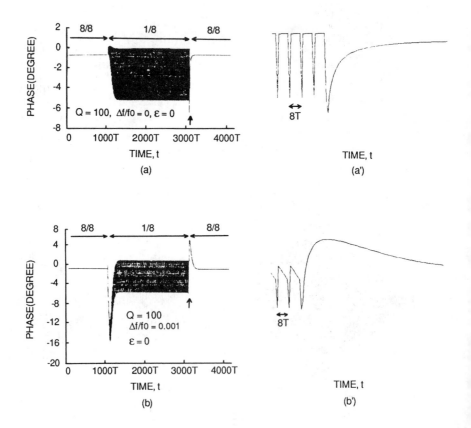

Figure 5.3 Examples of pattern-dependent jitter (patterns 1/8–8/8) for Q = 100: (a, a′) jitter due to finite Q and NZ pulse-width; (b, b′) jitter due to mistuning; and (c, c′) jitter due to waveform distortion.

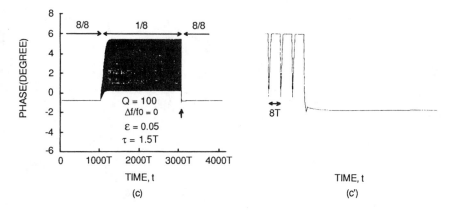

Figure 5.3 Continued

and waveform distortion. By comparing these figures, we see that jitter can be eliminated by making Q infinite, or making pulse width zero. The same pulse stream
pattern as that used for the analyses of Figure 5.2 is assumed. Pattern transients,
indicated by arrows, are expanded and shown in Figures 5.3(a′) and 5.4(a′). We can
see that spike-shaped phase deviations are caused every time pulse is applied to a
tuned circuit with the 1/8 pattern. Such jitter-fine structure can be attributed to finite
Q and nonzero pulsewidth.

Jitter generated by mistuning is analyzed in Figures 5.3(b) and (b′), and 5.4(b)
and (b′). Note that the influence of mistuning grows with larger Q values. It can be
shown that phase shift $\Delta\theta$ due to mistuning $\Delta f/f_0$ can be expressed as

$$\Delta\theta = 2Q\,\Delta f/f_0 \tag{5.1}$$

as discussed in Mayo [3]. Therefore $\Delta f/f_0$ for $Q = 500$ is five times smaller than
that for $Q = 100$.

The influence of waveform distortion is studied in Figures 5.3(c) and (c′), and
5.4(c) and (c′) in terms of a waveform having a single echo (details will be provided
below: see echo #2 in Figure 5.6(a) and assume $\varepsilon = 0.05$).

We can see in Figures 5.3 and 5.4 that the jitter caused by waveform distortion
is type B, and that others are type A (Problem 5.1). Amplitude-to-phase conversion
is also type B. As will be discussed in Chapter 6, the accumulation of type A jitter
is negligible as compared to that of type B. Therefore, we will concentrate on type
B, or narrow-sense systematic jitter, below.

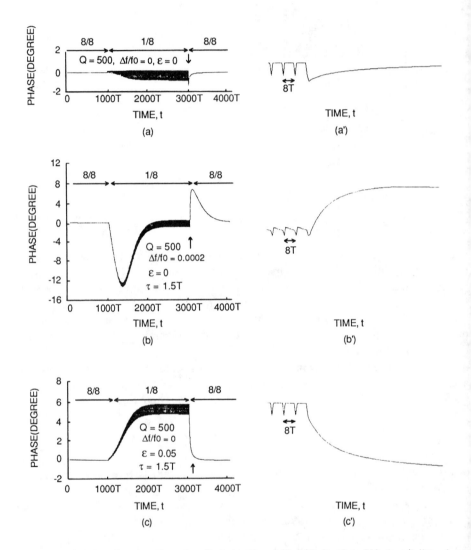

Figure 5.4 Examples of pattern-dependent jitter (patterns 1/8–8/8) for $Q = 500$: (a, a′) jitter due to finite Q and NZ pulse-width; (b, b′) jitter due to mistuning; and (c, c′) jitter due to waveform distortion.

5.2 JITTER GENERATION

We have suggested that dominant jitter in digital transmission design is due to waveform distortion and amplitude-to-phase conversion. We shall concentrate on waveform distortion here, since the behavior of amplitude-to-phase conversion is not yet completely clarified, but is supposedly almost the same as that of waveform distortion.

We can substantially eliminate jitter due to waveform distortion by employing a sufficiently high order of Bessel filtering. On the other hand, a filter with lower order is desirable from an economical standpoint. As a compromise, we accept some jitter degradation due to approximation by a lower-order filter, which will be studied in Section 5.2.1.

In addition to the approximation error mentioned above, we must also consider realization errors due to deviations in component values of filters from nominal design specifications. The influence of such realization errors is studied in Section 5.2.2.

5.2.1 Jitter due to Approximation Error

We suggested in Section 3.1 that Bessel filtering efficiently approximates waveform shaping that can simultaneously minimize ISI and jitter. Details are studied below to show that while NRZ systems are more efficient in minimizing jitter and ISI, RZ systems can be preferable from the standpoint of reducing crosstalk and the amplitude variation of the clock component.

In Section 4.3 we studied waveform function $\hat{H}(f)$, and how convenient it is to check the behavior of the clock component with it. A function $\tilde{H}(f) = H(f_0 - f)H(f)$ is useful for understanding the physical meaning of $\hat{H}(f)$. Therefore, both $\hat{H}(f)$ and $\tilde{H}(f)$ are shown in Figure 5.5 for the response of Bessel filters to RZ as well as NRZ pulses. The relation between $\hat{H}(f)$ and $\tilde{H}(f)$ can be expressed as

$$\hat{H}(f) = \sum_{n=-\infty}^{\infty} \tilde{H}(f + nf_0). \tag{5.2}$$

It should be easy to understand how $|\tilde{H}(f)|$ in Figure 5.5(a1') can be obtained from $|H(f)|$. Then we can refer to (5.2) to see how we can get $|\hat{H}(f)|$ in Figure 5.5 (a1') from $|\tilde{H}(f)|$. Note that $\tilde{H}(f)$ for $f < 0$ is not shown in the figure. Other figures in Figure 5.5 can be similarly understood.

We can see from Figures 5.5 (a2') and (b2') that maximum jitter decreases as order n of the filter increases, except for $N = 1$, 2 and 3. We can also see, by comparing Figures 5.5(a1') and (b1'), that amplitude variation due to pulse pattern change can be smaller with the response to RZ pulse than to NRZ pulse.

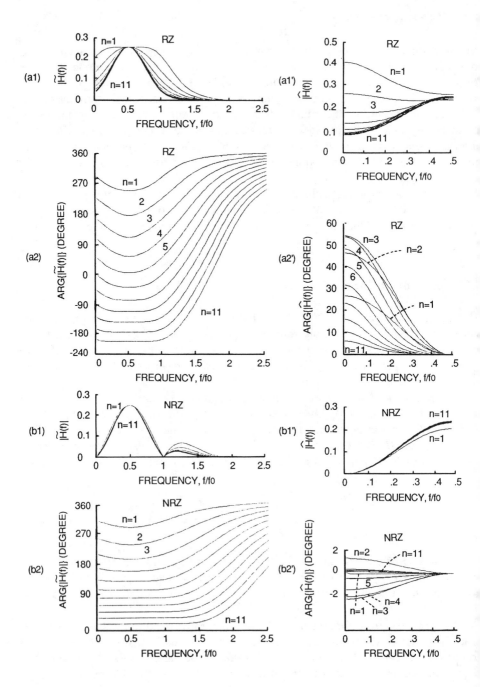

5.2.2 Jitter due to Realization Error

Inevitably, in practical environments, the component values of shaping filters deviate from specified values and cause shaping errors. We have studied in Section 3.3 that shaping errors can be evaluated in terms of echoes.

Let us first study simple cases with a single echo, as shown in Figure 5.6(a). The echo tends to move to the center of a timeslot when we design Gibby-Smith shaping to minimize ISI (Problem 5.2). Amplitude and phase deviations in the frequency domain that cause a single echo are shown in Figure 5.6(b).

Now, let us study the influence of an echo on a waveform function. Spectrum $H(f)$ of an even symmetric pulse of unit amplitude, followed by an echo with amplitude ε and delay τ can be expressed as

$$H(f) = H_0(f)\{1 + \varepsilon \exp(-j2\pi f\tau)\} \tag{5.3}$$

where $H_0(f)$ is the spectrum of the even symmetric pulse. Waveform function $\hat{H}(f)$ is given by

$$\hat{H}(f) = E(f, 0)\{1 + 2\eta(f, \tau)\varepsilon \exp(-j\pi\tau f_0) + \varepsilon^2 \exp(-j2\pi\tau f_0)\} \tag{5.4}$$

where

$$E(f, \tau) = \sum_{l=-\infty}^{\infty} H_0(f_0 - \overline{f + lf_0})H_0(f + lf_0) \cos\{2\pi\tau[(f_0/2) - \overline{f + lf_0}]\} \tag{5.5}$$

and

$$\eta(f, \tau) = E(f, \tau)/E(f, 0) \tag{5.6}$$

The angle of $\hat{H}(f)$ in (5.4) for $\varepsilon \ll 1$ can be expressed as

$$\arg\{\hat{H}(f)\} \cong -2\varepsilon\eta(f, \tau) \sin(\pi f_0\tau) \tag{5.7}$$

For a practical waveform (i.e., when $H(f)$ is very small for $|f| \geq f_0$),

$$\eta(f, \tau) \simeq \cos\{2\pi(f_0/2 - f)\tau\} \tag{5.8}$$

Figure 5.5 Auxiliary function $\hat{H}(f)$ and waveform function $\hat{H}(f)$ for the response of Bessel filters to RZ and NRZ pulses: (a1) amplitude of $\hat{H}(f)$ for RZ pulse response; (a2) angle of $\hat{H}(f)$ for RZ pulse response; (b1) amplitude of $\hat{H}(f)$ for NRZ pulse response; (b2) angle of $\hat{H}(f)$ for NRZ pulse response; (a1′) amplitude of waveform function for RZ pulse response; (a2′) angle of waveform function for RZ pulse response; (b1′) amplitude of waveform function for NRZ pulse response; and (b2′) angle of waveform function for NRZ pule response.

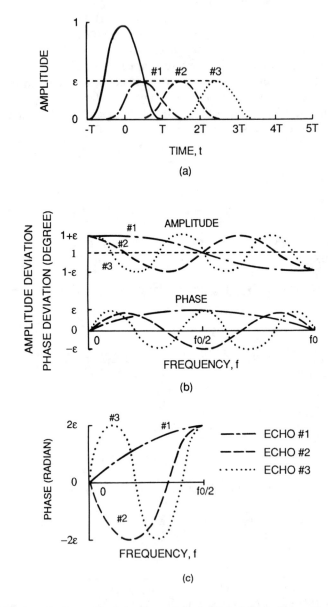

Figure 5.6 Influence of echo on the angle of waveform function: (a) waveform with a single echo (#1 #2 or #3), (b) amplitude and phase deviations in frequency responses due to a single echo and (c) deviations in the angle of waveform function due to a single echo.

Equation (5.7) is plotted in Figure 5.6(c) for echoes #1, #2, and #3. Figure 5.6(b) may suggest that a #2-type echo could be most probable in practical systems. Also, Figure 5.5(a2′) suggests that a #1-type echo is predominant in Bessel filtering (Problem 5.3). When the number of echoes is more than one, we can apply the law of superposition provided that $\varepsilon \ll 1$.

Jitter due to echo #2 is plotted by a solid line in Figure 5.7 for several representative repetitive pulse patterns. A real example is plotted by a dotted line for comparison. Sparse patterns (1/4, 1/8) and dense patterns (8/8, 7/8) in the real example could include the influence of amplitude-to-phase conversion. Also, the jitter ripple in the real example may be interpreted as the influence of the jitter-fine structures we studied in Figure 5.3.

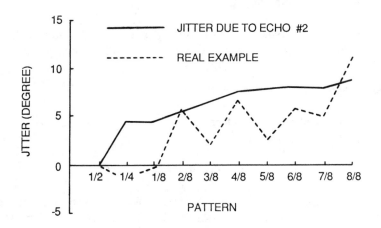

Figure 5.7 Examples of static pattern-dependent jitter due to echo #2 in Figure 5.6(a).

5.3 CONTROLLING JITTER GENERATION

5.3.1 Criteria for Jitter-Free Waveform Shaping

It should be apparent that the condition for jitter-free waveform shaping must be

$$\arg\{\hat{H}(f)\} = \text{constant},\ 0 \le f \le f_0/2 \tag{5.9}$$

Let us study (5.9) for waveforms with echoes, dealt with above. Equation (5.9) is satisfied by $\hat{H}(f)$ in (5.4) if

$$\eta(f, \tau) = \text{constant (independent of } f) \tag{5.10}$$

or

$$\tau f_0 = \text{integer} \qquad (5.11)$$

Practical waveforms do not usually satisfy (5.10). Therefore (5.11) must be satisfied. This is also true when the number of echoes is more than one. In other words, a waveform that does not generate jitter is given by a symmetrical wave followed by echoes with delay times that are integer multiples of pulse period T. This type of waveform will be defined below as a broad-sense symmetrical waveform. Note that this type of waveform shaping can increase ISI.

5.3.2 Optimizing Timing Path

We know that jitter-free shaping can adversely affect the reduction of ISI. In this section, separate shaping for timing and information paths is considered.

Broad-Sense Symmetrical Waveform Shaping

Let us consider one of the simplest broad-sense symmetrical shapings, $H_{bss}(f)$, that is, the one incorporating only a single echo

$$H_{bss}(f) = H_0(f)[1 + \varepsilon e^{-sT}] \qquad (5.12)$$

where $H_0(f)$ is a shaping function with a symmetrical response. We can use a trick here, that of replacing $H_0(f)$ with Bessel-filtering function $H_{BM}(f)$ of the Mth order, given by

$$H_{BM}(f) = \left[\sum_{i=0}^{M} b_M(i)(s/f_b)^i \right]^{-1} \qquad (5.13)$$

where f_b should be chosen to approximate a Nyquist or Gibby-Smith criterion (Problem 5.4). Values of $b_M(i)$ can be found in [4]. Then we can obtain

$$H_{bssM}(f) = \left[\left(\sum_{i=0}^{M} b_M(i)(s/f_b)^i \right)(1 - \varepsilon e^{-sT}) \right]^{-1} \qquad (5.14)$$

where it is assumed that $\varepsilon \ll 1$. Applying an approximate expansion for e^{-sT}, that is,

$$e^{-sT} = 1 + (-sT) + (-sT)^2/2! + \ldots \qquad (5.15)$$

yields

$$H_{bssMN}(f) = \left[\sum_{j=0}^{N} B_M(j)(s/f_b)^j \right]^{-1} \qquad (5.16)$$

where

$$B_M(j) = b_M(j) - [\varepsilon(-T)^j/j!] \sum_{k=0}^{j} k!\,_j C_k b_M(k)/(-T)^k \qquad (5.17)$$

and where $_i C_k$ stands for binomial co-efficient. A steepest-descent method can be applied to optimize ε in (5.17), leading to the minimization of worst-case jitter [5]. Choosing M to be $N + 1$ was found to be convenient for optimization.

A simpler, but less systematic, approach may be to apply a steepest-descent method directly to the Bessel filters in Figures B.1 and B.2. Taking the element values of the Bessel filters in Figure B.2 as initial element values, we can vary the value of each element according to steepest-descent law and obtain optimum values [6]. Some examples of RZ pulse responses and their waveform functions are shown in Figures 5.8(a) and (b), respectively.

Compared with the Bessel filter cases of the same order, the maximum jitter is greatly reduced. Response waveforms for order 7 and order 9 filters can be interpreted as even symmetric pulses followed by an echo of positive polarity delayed by T. The response of an order 11 filter incorporates echoes of negative and positive polarities delayed by T and $2T$, respectively.

Reducing Amplitude Variation of the Clock Component

One way to reduce amplitude-to-phase conversion is to reduce the amplitude variation of the extracted clock component with the proper choice of line coding. Another is to adopt separate waveform shaping specifically designed for the timing path. Vector analysis in terms of repetitive patterns is convenient for the study of the reduction of amplitude variation. Let us define pattern vector $\mathbf{W}_k = (W_1^k, W_2^k, \ldots, W_M^k)$ that corresponds to the kth digital pattern ($k = 1, 2, \ldots, K$, K: number of total patterns) and waveform vector $\hat{\mathbf{H}} = (\hat{H}_1, \hat{H}_2, \ldots, \hat{H}_M)$ where MT is a pattern period and

$$W_m^k = \left| W_k\left(\frac{m-1}{M} f_0 \right) \right|^2 \qquad (5.18)$$

$$\hat{H}_m = \hat{H}\left(\frac{m-1}{M} f_0 \right). \qquad (5.19)$$

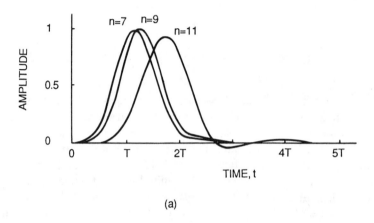

(a)

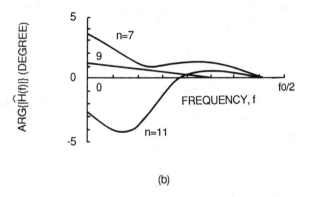

(b)

Figure 5.8 Broad-sense symmetrical waveform shaping; (a) broad-sense symmetrical waveforms obtained by modifying Bessel filtering functions, and (b) angle of waveform functions for broad-sense symmetrical waveforms.

Magnitude of a timing wave can be represented (apart from a constant) as

$$X_k^{(2)}(f_0) = \mathbf{W}_k \cdot \hat{H} \overset{\Delta}{=} |\mathbf{W}_k| \cdot |\hat{\mathbf{H}}| \cdot \cos \theta_k \qquad (5.20)$$

We want to minimize the variation ratio

$$R = \max_k \{|\mathbf{W}_k| \cos \theta_k\} / \min_l \{|\mathbf{W}_l| \cos \theta_l\} \qquad (5.21)$$

under the assumption that zero code is suppressed (i.e., at least one of $a_i s$ is not zero).

Equation (5.21) is illustrated in Figure 5.9. Note there that many of the pattern vectors (shown by dotted lines) can be eliminated, without the knowledge of waveform vector $\hat{H}$, from candidates that contribute extreme amplitudes (minimum and maximum).

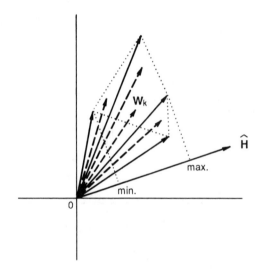

Figure 5.9 Vector analysis of amplitude variations in the clock component.

Optimum waveform functions that can minimize R are studied in [6]. A steepest-descent method was applied in obtaining optimum $\hat{H}$ in Figure 5.9. Typical results for optimum and simplified prefiltering are summarized in Table 5.2 for repetitive AMI sequences with an eight-bit period.

Table 5.2
Reduction of Amplitude Variation in the Clock Component,
in Terms of Prefiltering in the Timing Path

Prefiltering	Variation Ratio, R	Digital Patterns Giving Extreme Clock Amplitudes	
None	16	max	1 -1 1 -1 1 -1 1 -1
		min	1 0 0 0 0 0 0 0
Single Tap Transversal Filter	7.6	max	1 -1 0 1 -1 1 0 -1
		min	1 0 0 0 0 0 0 0
Optimum	2.3	max	1 -1 1 -1 1 -1 1 -1
		min	1 -1 0 0 0 0 0 0

5.3.3 Simultaneous Optimization of Jitter and ISI

Tail Shaping Plus Predistortion

In trying to minimize ISI, we sometimes get a waveform with an oscillatory tail, a
in Figure 5.10(a). In general, an oscillatory tail makes a waveform asymmetrica
and leads to considerable pattern-dependent jitter. Such oscillation can be reduce
by generating an echo as shown in Figure 5.10(b). This, however, entails som
increase in ISI. We can apply the predistortion (or nonlinear equalization [3]) tech
nique illustrated in Figure 5.11 to alleviate this problem. All we have to do is us
pulses with an undershoot as shown in Figure 5.11(a′).

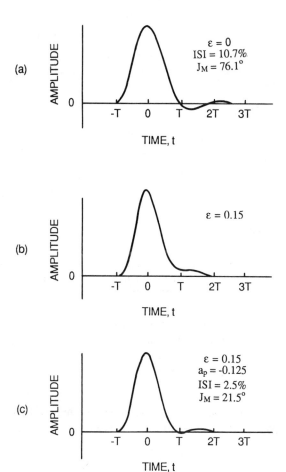

Figure 5.10 Tail-shaping plus predistortion: (a) shaped pulse with an oscillatory tail; (b) echo generation for reducing oscillation in the tail; and (c) echo cancellation by means of predistortion.

Waveforms in Figure 5.10 are obtained by applying shaping function $H_{BssMN}(f)$ in (5.16). Note that $T/2$ is substituted for T in (5.17) with $N = 5$. The results depend on the choice of $[b_M(i)]$, though detailed descriptions are not included here. For the examples in Figure 5.10, M is chosen to be 5 and 0 is substituted for the value of $b_5 (5)$.

The amplitude of the oscillatory tail is decreased as echo amplitude ε is increased (compare Figures 5.10(a) and (b)). The increase in ISI can be cancelled by the predistortion technique mentioned above. The efficiency of this ˙thod is apparent in Figure 5.10(c). Predistortion with an undershoot of $a_p = -0...5$ is applied

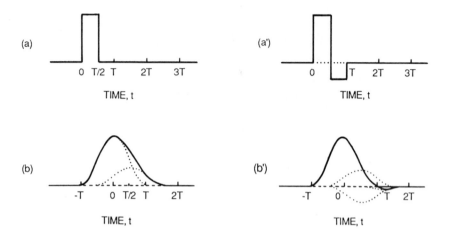

Figure 5.11 Principles of predistortion: (a) pulse with the RZ format; (b) shaped pulse with an echo generated half a timeslot after the main response; (a') pulse with predistortion; and (b') shaped pulse with an echo cancelled by predistortion.

to waveform (b) with a minimum oscillatory tail ($\varepsilon = 0.15$) to obtain waveform (c)[5]. We can see that ISI and maximum jitter (J_M) are reduced by a factor of 4.

Choosing Line Coding

As suggested above, jitter depends on the choice of line coding. Typical examples are shown in Figure 5.12 for the 5th-order Bessel filtering of an RZ pulse stream.

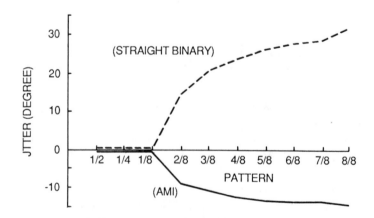

Figure 5.12 Dependence of static-pattern jitter on line coding.

AMI coding is more useful for reducing jitter than is the straight binary scheme (Problem 5.5).

PROBLEMS

Problem 5.1 Speculate on why jitters in Figures 5.3(c) and 5.4(c) are type B, and others are type A.

Problem 5.2 Explain why reducing ISI tends to cause echo to appear at the center of a timeslot when Gibby-Smith waveform shaping is employed (Figure 5.6).

Problem 5.3 Explain why echo #2 in Figure 5.6 is most likely to occur in practical systems, and echo #1 is predominant in Bessel-type waveform shaping.

Problem 5.4 Explain what "to approximate a Nyquist or Gibby-Smith criterion," means with regard to f_b in (5.13).

Problem 5.5 Explain why less jitter occurs with AMI coding compared to the straight binary scheme in Figure 5.12.

REFERENCES

1. Byrne C.J., *et al.*, Systematic Jitter in a Chain of Digital Regenerators, *Bell Syst. Tech. J.*, Vol. 42, November 1963, pp. 2679–2714.
2. Manley J.M., The Generation and Accumulation of Timing Noise in PCM Systems—An Experimental and Theoretical Study, *Bell Syst. Tech. J.*, Vol. 48, March 1969, pp. 541–613.
3. Mayo J.S., A Bipolar Repeater for Pulse Code Modulation Signals, *Bell Syst. Tech. J.*, Vol. 41, January 1962, pp. 25–97.
4. Storch L., Synthesis of Constant-Time Delay Ladder Networks Using Bessel Polynomials, *Proc. IRE*, Vol. 42, November 1954, pp. 1666–1675.
5. Takasaki Y., Optimizing Pulse Shaping for Baseband Digital Transmission with Self-Bit Synchronization, *IEEE Trans. Commun.* Vol. COM-28, No. 8, August 1980, pp. 1164–1172.
6. Takasaki Y., Timing Extraction in Baseband Pulse Transmission, *IEEE Trans. Commun.*, Vol. COM-20, No. 5, October 1972, pp. 877–884.

Chapter 6
Jitter Accumulation

We study Chapman's model for analyzing the accumulation of transitional, quasi-static, and dynamic, or rms, jitter. Only by using the results of static jitter measurements can rms jitter be estimated.

We will then modify Chapman's model to analyze the influence of the processing delay in respective repeaters in a chain. We will see that the behavior of jitter accumulation changes completely in the presence of processing delay: extra-accumulation is maximized for a ratio of Q to M (M = processing delay) of approximately 3.

Then, we concentrate on the accumulation of alignment jitter in terms of quasistatic analysis based on the modified Chapman's model. Accumulation of alignment jitter was negligible with the unmodified version of that model. With the modified version, we will see that alignment jitter can accumulate in proportion to the 0.6th to 0.7th power of N (N = number of repeaters in a chain).

After summarizing jitter classifications, we will study three methods for alleviating the extra-accumulation; the use of timing filters having considerable delay (i.e., transversal-type SAW filters), reducing the period of quasistatic patterns, and choosing the Q/M ratio much larger or much smaller than 3.

6.1 TIMING JITTER ACCUMULATION

Byrne, *et al.* have shown that the accumulation of pattern-dependent systematic jitter (pattern jitter) becomes predominent in the repeater chain [1]. This happens because, while systematic jitter accumulates in proportion to the number of repeaters N, random jitter follows square root N law in accumulation. Manley has shown that pattern jitter (broad-sense systematic jitter) is composed of type A jitter and type B (narrow-sense systematic) jitter [2]. While type B jitter accumulates in proportion to N, the accumulation of type A jitter is almost negligible, as we shall discuss below. Here, let us restrict our discussion to type B jitter.

Chapman's model, shown in Figure 6.1, has proved useful in analyzing jitter accumulation in a repeater chain [1]. Several assumptions have been made to make the model tractable. Namely: the same jitter is injected at each repeater; jitter adds linearly from repeater to repeater; and the timing tank is a simple single-tuned circuit.

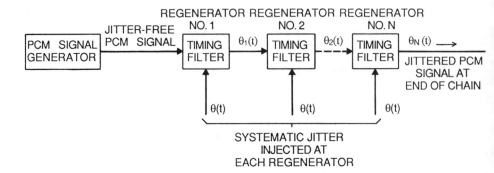

Figure 6.1 Chapman's model for analyzing jitter accumulation. *Source:* Byrne C.J., *et al.*, "Systematic Jitter in a Chain of Digital Regenerators," *Bell Syst. Tech. J.*, Vol. 42, No. 6, November 1963, p. 2681. Reprinted with permission. Copyright© 1963 AT&T.

As far as phase modulation is concerned, the timing tank is equivalent to a single-pole low-pass filter. The pole corresponds to the half-bandwidth of the tuned circuit. Therefore,

$$\Theta_o = \frac{1}{1 + (s/B)} \Theta_i \tag{6.1}$$

where Θ_i and Θ_o are the transforms of input and output jitter, respectively, and $B = \omega_0/2Q$ is the half-bandwidth of the tank.

Since the jitter introduced in each repeater $\theta(t)$ is assumed to be identical, we can express the jitter at the end of a chain of N repeaters as

$$\Theta_N(s) = \sum_{n=1}^{N} \Theta(s)\left(\frac{1}{1 + (s/B)}\right)^n$$

$$= \Theta(s)\frac{B}{s}\left[1 - \left(\frac{1}{1 + (s/B)}\right)^N\right] \tag{6.2}$$

where $\Theta(s)$ is the transform of jitter $\theta(t)$ introduced at each repeater.

The square of the magnitude of the transfer function given in (6.2) is plotted in Figure 6.2 for a range of values of N. We can use these curves to study the

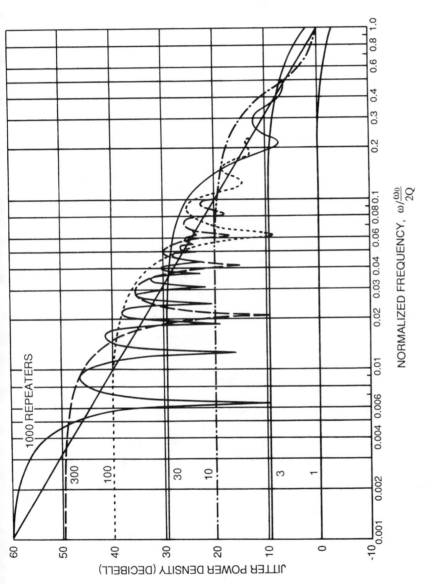

Figure 6.2 Jitter spectrum due to a random pattern (computed from the model). *Source:* Byrne C.J., *et al.*, "Systematic Jitter in a Chain of Digital Regenerators," *Bell Syst. Tech. J.*, Vol. 42, No. 6, November 1963, p. 2685. Reprinted with permission. Copyright© 1963 AT&T.

accumulation of the power density of jitter in a chain of repeaters [1], and see that jitter accumulates only at very low frequencies.

Now, let us study the estimation of $\Theta(s)$, the jitter introduced at each repeater, which is elaborated on in [1]. Consider a sample from a bit sequence shown in Figure 6.3(a). It is assumed that the jitter introduced during the time span of a particular

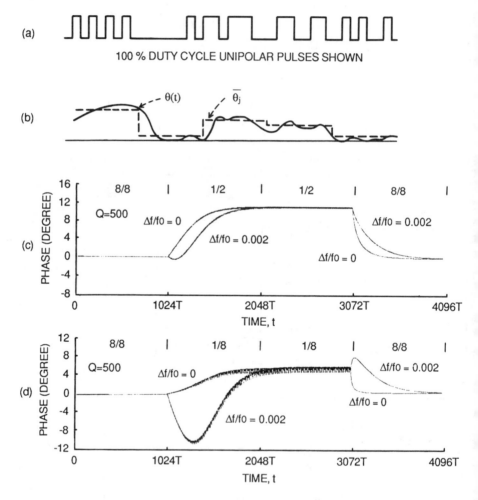

Figure 6.3 Estimation of $\Theta(s)$ by means of static jitter measurement: (a) sample from a bit sequence; (b) actual and approximate jitter waveforms; *Source:* Byrne C.J., *et al.*, "Systematic Jitter in a Chain of Digital Regenerators," *Bell Syst. Tech. J.*, Vol. 42, No. 6, November 1963, p. 2688. Reprinted with permission. Copyright© 1963 AT&T. (c) combined type A (due to mistuning) and type B (due to waveform distortion) jitter for a pattern combination of 1/2–8/8; and (d) combined type A (due to mistuning) and type B (due to waveform distortion) jitter for a pattern combination of 1/8–8/8.

block will be caused chiefly by the bits within the block, and only slightly influenced by bits outside the block.

Therefore, it is reasonable to assign a particular value to the phase shift of each block. The resultant approximate jitter waveform is shown in Figure 6.2(b) by dotted lines. The value of phase shift $\bar{\theta}_j$, for a particular block with a pattern j, can be determined by repeating the block in a sufficiently long period and measuring the resulting dc phase shift.

We have to be especially careful when the value of Q is large, since it can be difficult in the measurement to eliminate type A jitter. The reason we have to eliminate type A jitter is explained in Figure 6.4. In the figure, (a) and (b) illustrate accumulation of type A and type B jitter, respectively [2]. Note that type A jitter does not include jitter components at very low frequencies, as opposed to type B. Although the amounts of type A and type B jitter are almost equal for $N = 1$ ($N =$ number of repeater in a chain), type A jitter stops accumulating at $N = 5$ or 6, as opposed to type B, which can accumulate limitlessly with large values of N (Problem 6.1). Since the measurement for $\bar{\theta}_j$ is done for $N = 1$, it is very important to eliminate the influence of type A components.

Examples of $\bar{\theta}_j$ measurement are shown in Figures 6.3(c) and (d) with $Q = 500$. Type B jitter is caused by echo #2 in Figure 5.6, with $\varepsilon = 0.05$. The jitter waveforms of Figures 6.3 (c) and (d) show the influence of type A jitter due to mistuning $\Delta f/f_0$ for pattern combinations of 8/8–1/8 and 8/8–1/2. Note that a pattern combination of 8/8–1/8 requires a longer period of repetition of the block than an 8/8–1/2 combination for eliminating the influence of type A jitter due to mistuning (Problems 6.2 and 6.3).

We have studied how to evaluate $\bar{\theta}_j$. Two questions follow: What is the reasonable block length for each block of the jitter waveform in Figure 6.3(b)? And how we can estimate $\Theta(s)$ in (6.2) by using measured values of $\bar{\theta}_j$s?

Each block should be as long as possible to minimize end effects, but also sufficiently short to allow the practical measurement of the required number of distinct patterns. As a compromise, the length 8 was chosen [1].

This method has proven very useful, because we can estimate rms jitter by using data obtained from the dc measurement of jitter for only 35 patterns [1].

It is shown in [1] that $\Theta(s)$, the spectral power density of $\theta(t)$, is flat at low frequencies and that the value of the density Φ is

$$\Phi = 2T_B\sigma^2 \tag{6.3}$$

where T_B is the time duration of each block, σ is the standard deviation of the block phase shifts $\bar{\theta}_j$, and

$$\sigma^2 = \frac{1}{M}\sum_j p_j\left(\bar{\theta}_j - \frac{1}{M}\sum_j p_j\bar{\theta}_j\right)^2 \tag{6.4}$$

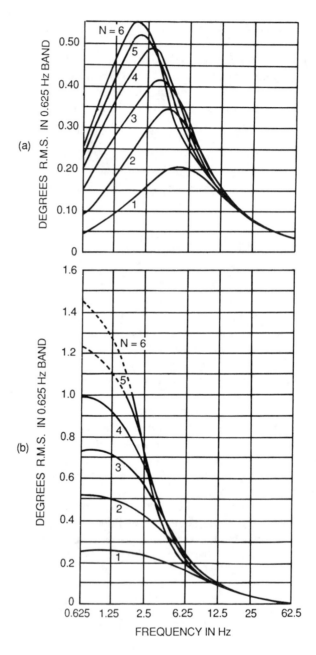

Figure 6.4 Accumulation of type A and type B jitter; (a) type A, and (b) type B. *Source:* Manley J.M., "The Generation and Accumulation of Timing Noise in PCM Systems—An Experimental and Theoretical Study," *Bell Syst. Tech. J.,* Vol. 48, March 1969, pp. 541–613. Reprinted with permission. Copyright© 1969 AT&T.

where M is the total number of distinct patterns and p_j is the probability of pattern j. It is also shown in [1] that mean square jitter $\overline{\theta_N^2}$ can be given by

$$\overline{\theta_N^2} \simeq \frac{1}{2} \Phi BN \tag{6.5}$$

for chains of more than 100 repeaters, where B and N are the half-bandwidth of the timing tank, and the number of repeaters in a chain, respectively. Theory and experiment were found to be in agreement, as shown in Figure 6.5.

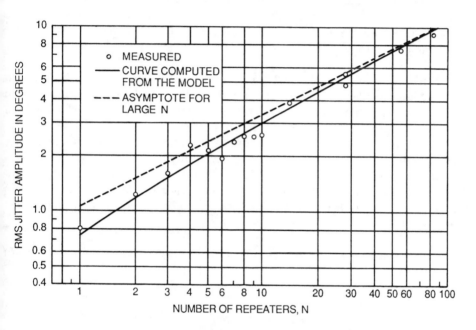

Figure 6.5 Root mean square jitter due to a random pattern versus the number of repeaters in the chain, measured and calculated. *Source:* Byrne C.J., *et al.,* "Systematic Jitter in a Chain of Digital Regenerators," *Bell Syst. Tech. J.,* Vol. 42, No. 6, November 1963, p. 2702. Reprinted with permission. Copyright© 1963 AT&T.

6.2 ALIGNMENT JITTER ACCUMULATION

While timing jitter affects the quality of received signals, alignment jitter consumes time crosshair budgets. Accumulation of alignment jitter is negligible in ordinary circumstances [2]. However, extra-accumulation can arise when processing delay is incorporated in respective repeaters in a chain. The extra-accumulation can become fatal (as was previewed in Section 1.6) when one long, repetitive pulse pattern is followed by another [3] (worst-case alignment jitter). Evaluating jitter based on the worst-case criterion, however, can lead to excessive requirements. On the other hand, the succession of identical patterns has an upper limit in ordinary systems. Therefore, we use the quasistatic criterion discussed in Section 1.6 to analyze alignment jitter accumulation.

6.2.1 Influence of Processing Delay

A model of a regenerative repeater that incorporates a processing delay of M time-slots is shown in Figure 6.6(a). In this case, we can no longer assume that jitter introduced at respective repeaters is identical. We must modify Chapman's model as shown in Figure 6.6(b). We can understand why jitter introduced at the Nth repeater can be expressed as $\theta(t - (N - 1)MT)$, with the help of Figure 6.6(c) (Problem 6.4).

Now, let us examine why extra-accumulation of alignment jitter arises due to delay difference $(N - 1)MT$. We can use simplified waveforms, shown in Figure 6.7(a), for the qualitative analysis of extra-accumulation. In the figure, N is assumed to be 2. Incoming jitter $\theta_{N-1}(t)$ and generated jitter $\theta(t - MT)$ is added at the retiming circuit in Figure 6.6(a), and applied to low-pass filter $H(j\omega)$ to result in output jitter $\theta_N(t)$. The difference $\theta_N(t) - \theta_{N-1}(t)$ is defined as alignment jitter $\theta_{aN}(t)$. The spike-shaped waveforms in $\theta_{aN}(t)$ are extra-accumulations of alignment jitter. Note that these spikes are caused by the difference of jitter slopes before and after low-pass filtering. The relation between $\theta_N(t)$ and $\theta_{N-1}(t)$ can be given by using their transforms $\theta_N(j\omega)$ and $\theta_{N-1}(j\omega)$, as

$$\theta_N(j\omega) = [\theta_{N-1}(j\omega) + \theta(j\omega) \exp(-j\omega(N - 1)MT)]H(j\omega) \qquad (6.6)$$

where

$$H(j\omega) = 1/(1 + j\omega/B), \quad B = \omega_0/2Q \qquad (6.7)$$

We can obtain a straightforward expression by further manipulation of (6.6)

$$\theta_N(j\omega) = \theta(j\omega) \exp(-j\omega NMT) \sum_{i=1}^{N} [H(j\omega) \exp(j\omega MT)]^i \qquad (6.8)$$

Equation (6.8) was used to calculate $\theta_N(t)$ and $\theta_{aN}(t)$ for various values of M.

Some results of calculations for $Q = 100$ and $N = 20$ are shown in Figures 6.7(b) and (c). We can see the rising and falling slopes of timing jitter $\theta_N(t)$ change, depending on the value of M. The steepest slopes are encountered for an M value of 30 in this example, seen, in Figure 6.7(c), to cause the largest extra-alignment jitter. Note also that extra-accumulation can be shown to be relatively small when

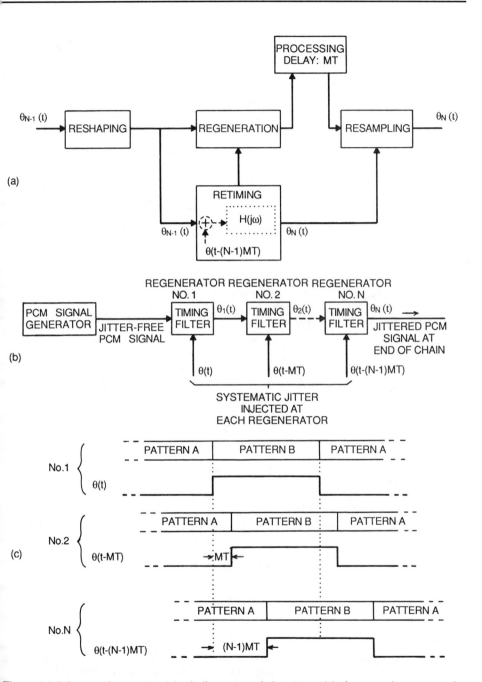

Figure 6.6 Influence of processing delay in jitter accumulation: (a) model of regenerative repeaters that incorporate processing delay; (b) modified Chapman's model; and (c) analysis of delay in jitter generation.

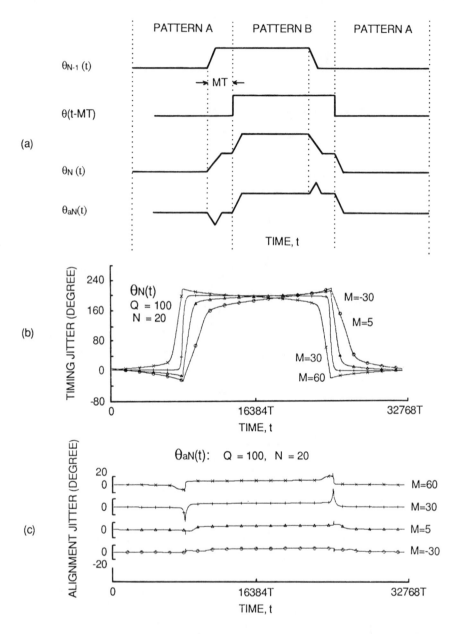

Figure 6.7 Analysis of extra-alignment jitter accumulation due to processing delay: (a) principles of extra-alignment jitter accumulation; (b) timing jitter accumulation for $N = 20$ and $Q = 100$ where M is varied as a parameter; and (c) alignment jitter accumulation for $N = 20$ and $Q = 100$ where M is varied as a parameter.

the value of *M* assumes negative values (Problem 6.5). The value of *M* can be negative when we use timing filters such as transversal-type SAW filters, which can incorporate several hundred timeslots of delay. We mainly study cases where *M* values are positive, because, depending on design requirements, SAW filters may or may not be applicable.

The dependence of worst-case alignment jitter on *M* values is shown in Figure 6.8. Figures 6.8(a) and (b) are calculated for quasistatic pattern period T_p of 32768T

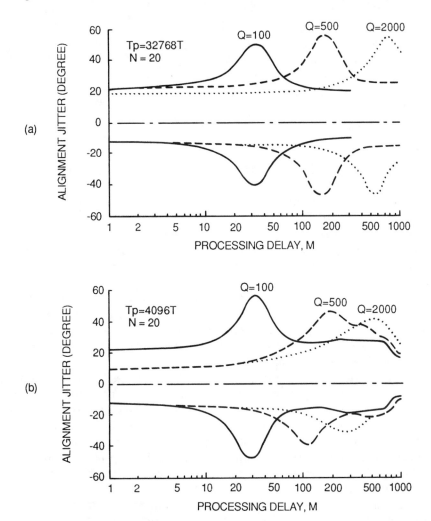

Figure 6.8 Dependence of worst-case alignment jitter on *M* values for $N = 20$; (a) $T_p = 32768T$ and (b) $T_p = 4096T$.

and 4096T, respectively. We can see the M values that maximize extra-accumulation are approximately proportional to the value of Q when T_p is sufficiently large. The choice of T_p depends on applications (Problem 6.6). Although it is not shown in the figure, the accumulation of extra-alignment jitter for negative M values is usually not large, but can become two to three times larger than that for $M = 0$.

6.2.2 Accumulation in a Chain of Repeaters

Now let us study accumulation of timing and alignment jitter based on the quasistatic criterion. Quasistatic pattern period T_p of 32768T and 4096T are used to see jitter accumulation in a chain of 100 repeaters.

Results for the processing delay of $M = 0$ are shown in Figure 6.9. We can see that the accumulation of timing jitter is small when T_p is small, due to the use of quasistatic analysis instead of static analysis. Also, accumulation tends to saturate even when T_p is large. Accumulation of alignment jitter is small even when T_p is made large.

Results for processing delay M that maximize extra-accumulation are shown in Figures 6.10, 11 and 12, for the Q values of 100, 500 and 2,000, respectively. Note that the saturation of timing jitter become less prominent. Also note that significant accumulation of alignment jitter is encountered.

We can expect that quasistatic jitter is almost equal to static (worst-case) jitter with $Q = 100$ and $Q = 500$ when T_p is as large as 32768T. On the other hand, quasistatic jitter is no longer equal to static jitter when T_p is equal to or smaller than 4096T.

Accumulation of timing and alignment jitter is summarized in Figure 6.13 for T_p of both 32768T and 4096T. Quasistatic analysis with square wave $\theta(t)$ having an amplitude of $10°$ is assumed—an assumption that is made nearly universally in this book. We can see that the behavior of jitter is drastically changed in the presence of processing delay in each repeater in a chain.

Note the difference between quasistatic analysis and static (worst-case) analysis. Timing and alignment jitter are seen to accumulate in proportion to N and to the 0.6–0.7th power of N, respectively, with regard to static analysis. On the other hand, with quasistatic analysis, the accumulation of quasistatic jitter slows for large N, and tends to decrease or stay constant as N becomes larger.

Jitter classifications are summarized in Table 6.1 based on the above discussions. Type B' is newly defined and pertains to jitter that exhibits extra-accumulation due to processing delay.

Note that jitter analysis based on an rms (dynamic jitter) criterion is convenient for specifying or evaluating the quality of received signals [4]. However, the rms criterion can be unreliable when evaluating time crosshair degradation, even for scrambled pulse sequences.

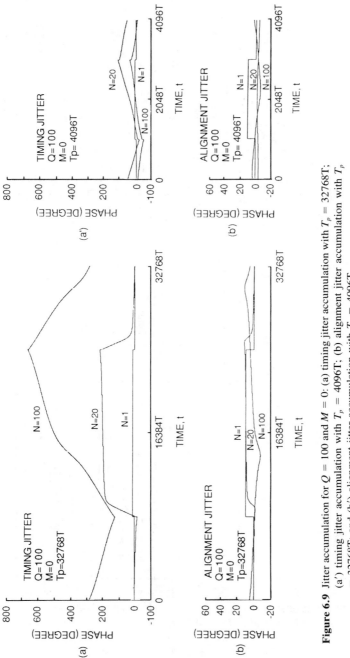

Figure 6.9 Jitter accumulation for $Q = 100$ and $M = 0$: (a) timing jitter accumulation with $T_p = 32768T$; (a') timing jitter accumulation with $T_p = 4096T$; (b) alignment jitter accumulation with $T_p = 32768T$; and (b') alignment jitter accumulation with $T_p = 4096T$.

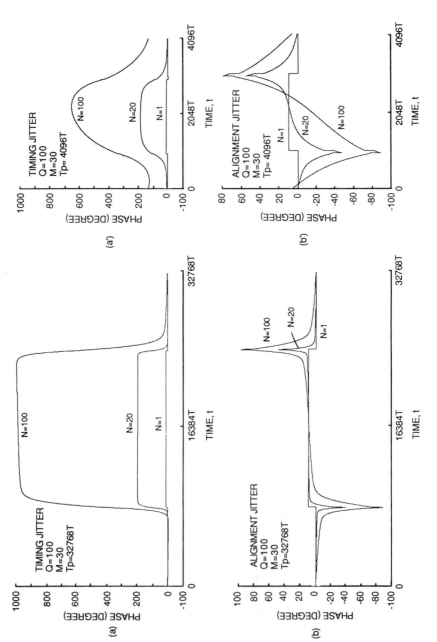

Figure 6.10 Jitter accumulation for $Q = 100$ and $M = 30$: (a) timing jitter accumulation with $T_p = 32768T$; (a') timing jitter accumulation with $T_p = 4096T$; (b) alignment jitter accumulation with $T_p = 32768T$; and (b') alignment jitter accumulation with $T_p = 4096T$.

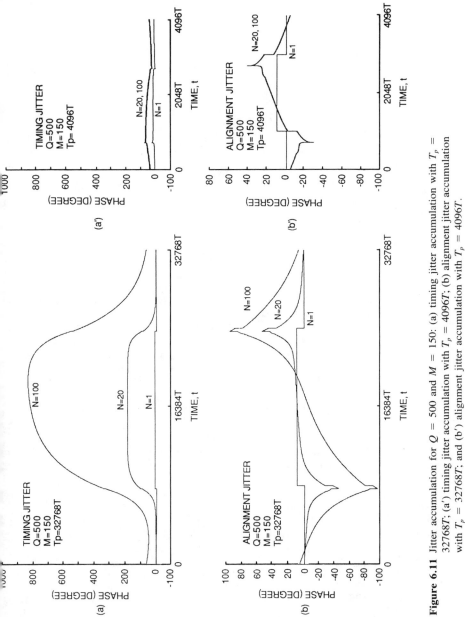

Figure 6.11 Jitter accumulation for $Q = 500$ and $M = 150$: (a) timing jitter accumulation with $T_p = 32768T$; (a') timing jitter accumulation with $T_p = 4096T$; (b) alignment jitter accumulation with $T_p = 32768T$; and (b') alignment jitter accumulation with $T_p = 4096T$.

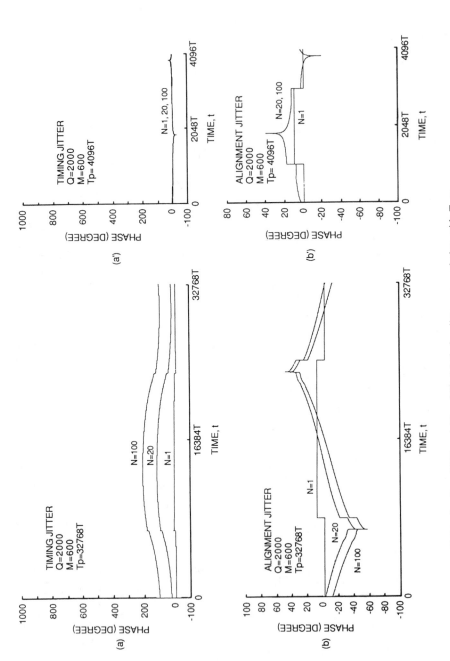

Figure 6.12 Jitter accumulation for $Q = 2000$ and $M = 600$: (a) timing jitter accumulation with $T_p = 32768T$; (a') timing jitter accumulation with $T_p = 4096T$; (b) alignment jitter accumulation

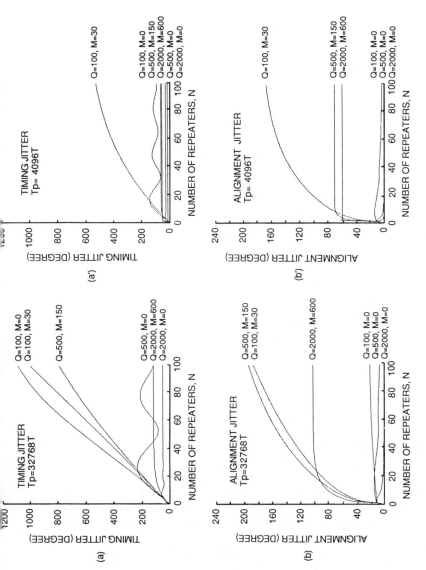

Figure 6.13 Summary of timing jitter and alignment jitter accumulation: (a) timing jitter accumulation for $T_p = 32768T$; (a') timing jitter accumulation for $T_p = 4096T$; (b) alignment jitter accumulation for $T_p = 32768T$; and (b') alignment jitter accumulation for $T_p = 4096T$.

Table 6.1

Summary of Jitter Classifications

Classification by Time Reference
    (a) Timing Jitter (Absolute Jitter)
    (b) Alignment Jitter
    (c) Spacing Jitter

Classification by Jitter Source
    (a) Noise Dependent Jitter (Random Jitter)
    (b) Pattern (Dependent) Jitter (Broad Sense Systematic Jitter)
        (i) Type A
        (i i) Type B (Narrow Sense Systematic Jitter)
        (iii) Type B' (Extra-Systematic Jitter)

Classification by Jitter Characteristics
    (a) Random Jitter
        (i) Noise Dependent
        (i i) Pattern Dependent (Type A)
    (b) Systematic Jitter (Type B, Type B')

Classification by Digital Sequence
    (a) Static Jitter
    (b) Transitional Jitter
    (c) Quasi-Static Jitter
    (d) Dynamic Jitter (rms Jitter)

Others
    (a) Jitter due to pulse stuffing
    (b) Jitter due to Digital PLL

Therefore, the use of static or transitional criterion can become necessary. It is convenient to use a square-wave pattern change with a sufficiently long period to measure static (pattern) jitter (see Figures 6.3(c) and (d)). Transitional jitter can also be measured with the same digital pattern. Static jitter and the peak-to-peak value of transitional jitter can be almost equal in some cases (see Figures 6.3(c) and 6.9(a) and (b) for $N = 1$ and 20, and 6.10(a)), but can be significantly different in others (see Figures 6.3(d) and 6.10(b)).

Note that the use of the static or transitional criterion can lead to excessive requirements or specifications in the design of digital transmission systems (see Figure 6.11 for an example). Figure 6.11(a) shows that timing jitter can become around 800° with $N = 100$. On the other hand, when we consider the design of a system where the maximum pattern period cannot exceed 4096 timeslots, it is seen in Figure 6.11(a') that timing jitter is less than 100°. Also, while alignment jitter is $\pm100°$,

ccording to worst-case design based on transitional criterion (Figure 6.11(b)), a
pecification of ±40° can be satisfied when practical conditions are taken into con-
ideration (Figure 6.11(b')).

We define a digital pattern, one with a period of square-wave wise pattern
hange that has been selected based on the requirements of practical system design,
s "quasistatic pattern." Jitter caused by a quasistatic pattern or the peak-to-peak
alue of such jitter is defined as "quasistatic jitter."

Jitter can also be generated by digital processing like pulse-stuffing and digital
PLL. Such jitter is beyond the scope of this book. Stuffing jitter (or waiting-time
itter) is elaborated on in [4].

.2.3 Alleviating Accumulation

We have seen that extra-accumulation of alignment jitter due to processing delay
an become unacceptably large from the standpoint of time crosshair design. We
an employ the three following methods to alleviate such extra-accumulation: the
se of a timing filter with a large delay such as a transversal-type SAW filter; the
ombined use of large Q and small T_p (pattern period); and the use of a Q/M ratio
nuch smaller (or larger) than 3.

The first method is considered the most efficient. If it is unacceptable due to
lesign requirements, we must resort to one of the other methods.

The applicability of the second method is studied in Figure 6.14. We can see
hat a time crosshair budget of 30° to 40° will have to be assigned for a pattern period

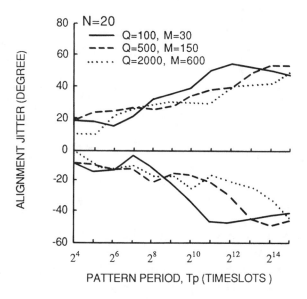

Figure 6.14 Dependence of extra-alignment jitter accumulation on $Q-T_p$ combinations.

constraint of less than 1000 timeslots. Note that calculations are made for $N = 20$, but note also, from Figure 6.13, that jitter should be kept constant for larger N values provided that T_p is sufficiently small.

The third method also requires a time crosshair budget of 30° to 40°, as seen from Figure 6.15. Note that the amount of jitter oscillates depending on the value of N, which is the case when the Q/M ratio deviates from around 3, (also seen from Figure 6.13).

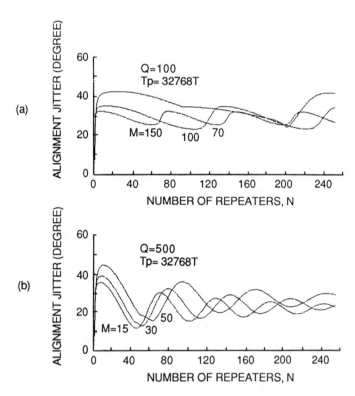

Figure 6.15 Dependence of extra-alignment jitter accumulation on Q/M ratios; (a) $Q = 100$ and (b) $Q = 500$.

PROBLEMS

Problem 6.1 Speculate on why type A jitter stops accumulating at a small value of N (N = number of repeaters in a chain).

Problem 6.2 Speculate on why a pattern combination of 1/8–8/8 generates smaller type B jitter than that of 1/2–8/8 in Figure 6.3.

Problem 6.3 Speculate on why timing jitter for a pattern combination of $1/8$–$8/8$ is more susceptible to mistuning than that for a combination $1/2$–$8/8$ in Figure 6.3.

Problem 6.4 Explain why jitter introduced at the Nth repeater can be expressed as $\theta(t - (N - 1)MT)$ in Figure 6.6.

Problem 6.5 Explain why extra-accumulation of alignment jitter is relatively small for negative M values.

Problem 6.6 Explain how the choice of T_p can depend on applications.

Problem 6.7 Explain the differences between static jitter, transitional jitter and quasistatic jitter in Table 6.1.

REFERENCES

Byrne, C.J. *et al.*, Systematic Jitter in a Chain of Digital Regenerators, *Bell Syst. Tech. J.*, Vol. 42, Nov. 1963, pp. 2679–2714.

Manley J.M., The Generation and Accumulation of Timing Noise in PCM Systems—An Experimental and Theoretical Study, *Bell Syst. Tech. J.*, Vol. 48, March 1969, pp. 541–613.

Takasaki Y., Alignment Jitter Accumulation in a Chain of Processing Node Regenerators, *Trans. IEICE*, Japan, Vol. E73, No. 10, October 1990, pp. 1712–1716.

Trischitta P.R., and E.L. Varma, Jitter in Digital Transmission Systems, Norwood, MA: Artech House, 1989.

Chapter 7

Computer Simulation

We study computer simulation to check the reliability of theoretical predictions based on the modified Chapman's model we discussed in Chpater 6. The fast Fourier transform (FFT) and large memory capacities help simulate real systems very closely.

We use quasistatic pulse streams, studied in the previous chapter, with a period of 4096 timeslots. We apply 32 samples per timeslot with interpolation to guarantee sufficient accuracy in jitter measurement. We also make use of a single echo with waveform reshaping for generating type B jitter, which will be introduced at respective repeaters in the simulated chain.

Then, we conduct a computer simulation to find that theoretical predictions are reliable under practical conditions: jitter changes sufficiently slowly or high frequency jitter is relatively small to comply equivalently with the second assumption of Chapman's model. We also discuss causes that make theoretical prediction deviate from the results of simulation. Such causes include nonzero rise and fall times of quasistatic jitter, and jitter fine structure.

Finally, we simulate the methods for alleviating extra-jitter accumulation discussed in Section 6.2.3 to confirm the reliability of theoretical predictions.

7.1 SIMULATION PROGRAM

The simulation program is designed to simulate the real world as closely as possible, and does so with the help of the FFT, large memory capacities, and waveform interpolation. The program list is shown in Table 7.1. First, we define the length of a quasistatic pulse sequence, NPRD. In this example a length of 4096 timeslots is used, indicated as NPRD = 4096 in the table. We also define the number of samples per timeslot, NSMP. NSMP is 32, so sufficient precision is obtained in calculating jitter with the help of waveform interpolation. Therefore, 131,072 samples are used for the calculation (NS = NSMP * NPRD).

Table 7.1

Computer Simulation Program

```
PROGRAM JITTER

DIMENSION RE(2,131072), AI(2,131072)
DIMENSION JTR(4096), CLK(4096)

NPRD=4096
NSMP=32
NS   =NSMP*NPRD

C PARAMETERS
        Q    =100
        DF   =0.0
        AECHO=0.05
        DECHO=1.5
        M    =30

C PATTERN JP=1: 1/8-8/8, JP=2: 2/8-8/8, ...
C         JP=8: 1/4-8/8, JP=9: 1/2-8/8
        JP=9

        NR=100
     DO 100 N=1,NR
        CALL CLCK(CLK, JTR, M)
        CALL PTTRN(RE, CLK, JP)
                    CALL FFT(RE, AI)
        CALL SPCTR(RE, AI, AECHO, DECHO)
                    CALL IFFT(RE, AI)
        CALL RCTFY(RE)
                    CALL FFT(RE, AI)
        CALL TANK(RE, AI, Q, DF)
                    CALL IFFT(RE, AI)
        CALL JITTER(N, RE, JTR, CLK)
100     CONTINUE

        STOP
        END
```

Next, we define parameters for simulation. Timing tank Q and mistuning $\Delta f/f_0$ are defined by Q and DF, respectively. We generate type B jitter with echo #2, shown in Figure 5.6. Normalized echo amplitude, and delay are defined by AECHO and DECHO, respectively. Processing delay is also defined by M. DECHO and M are expressed in timeslots.

Then, we define patterns of quasistatic pulse sequences for simulation in terms of pattern number JP. For example, JP = 1 defines a sequence of 1/8 pattern followed by another sequence of 8/8 pattern, each with a length of 2048 timeslots. This type of pattern is indicated as 1/8–8/8. JP = 9 defines a 1/2–8/8 sequence.

Now we can start simulation by utilizing seven types of subroutines listed in the table. We can understand the functions of these subroutines with the help of Figure 7.1. Subroutine CLCK generates clock pulses for pattern generator PTTRN.

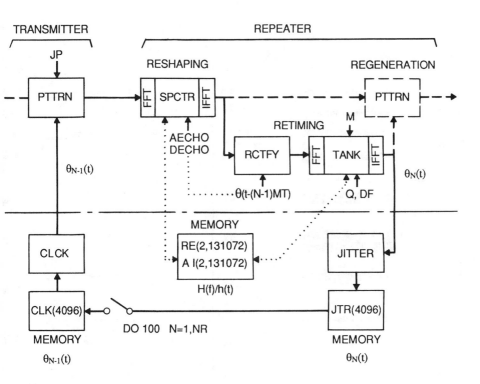

Figure 7.1 Block diagram of computer simulation program.

Subroutine PTTRN generates a quasistatic AMI sequence in accordance with parameter JP.

The generated sequence is applied to the reshaper that comprises FFT, reshaping function SPCTR, and inverse fast Fourier transform (IFFT). Subroutine SPCTR generates a main response and its echo. The main response is an impulse response of the raised cosine filter that satisfies Nyquist's criterion. The echo is defined by parameters AECHO and DECHO, as mentioned above.

The quasistatic AMI sequence is applied to the retiming circuit that comprises rectifying function RCTFY and narrow-band single-tuned filtering function TANK. Subroutine RCTFY carries out full-wave rectification and clipping as shown in Figure 4.3. The single-tuned filtering is defined by parameters Q and DF as mentioned above. Note that a delay of $-M$ timeslot is applied to the extracted clock component to simulate the processing delay.

Subroutine JITTER measures timing jitter $\theta_N(t)$ and calculates alignment jitter. The results are stored in memory JTR. These data are transferred to memory CLK in the next cycle and applied to subroutine CLCK, which in turn drives subroutine

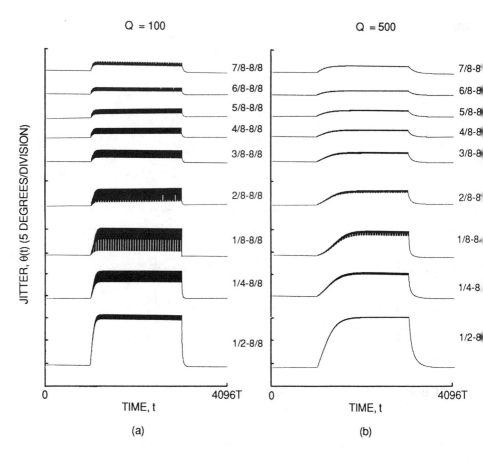

Figure 7.2 Quasistatic jitter $\theta(t)$ for (a) $Q = 100$, (b) $Q = 500$, and (c) $Q = 2000$.

PTTRN to generate a jittered pulse stream. Thus, jitter accumulation in a chain of repeaters can be simulated.

Figure 7.2 shows $\theta(t) = \theta_1(t)$ taking timing tank Q and the pattern of quasistatic pulse sequence as parameters. Other parameters, such as mistuning $\Delta f / f_0$, echo amplitude ε and echo delay τ are supposed to assume the values shown in Table 7.1. As a consequence, jitter waveforms in Figure 7.2 consist mainly of type B jitter. We can see that a quasistatic pattern of $1/2$–$8/8$ contributes the largest amount of jitter (Problem 7.1). Therefore, we mainly use this pattern for our study below. That is, we mainly consider a quasistatic jitter, with a period of 4096 timeslots and peak-to-peak value of around $10°$, introduced at each repeater; this is convenient when comparing results with theoretical prediction.

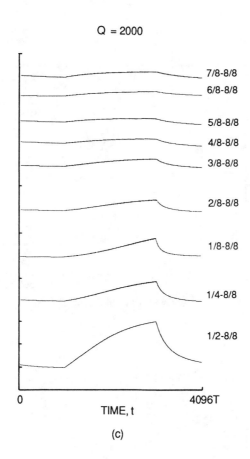

Figure 7.2 Continued.

7.2 RESULTS OF SIMULATION

7.2.1 Influence of Processing Delay and Patterns

Figure 7.3 shows the dependence of extra-alignment jitter accumulation on the value of processing delay M for two types of quasistatic patterns, and for $N = 20$. Comparison with the theoretical prediction shown in Figure 6.8 reveals that the values of M where maximum jitter is encountered are in good agreement, especially where pattern 1/2–8/8 is concerned.

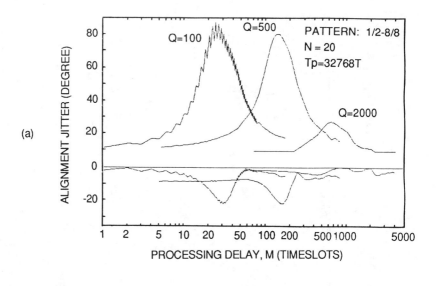

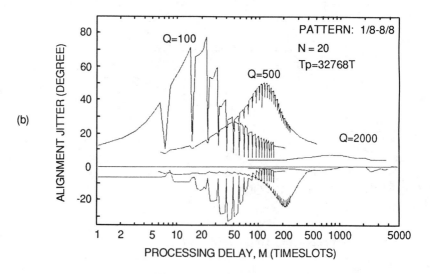

Figure 7.3 Dependence of extra-alignment jitter accumulation on processing delay M; (a) extra-alignment jitter for quasistatic pattern 1/2–8/8, and (b) extra-alignment jitter for quasistatic pattern 1/8–8/8.

Notable differences are saw-toothed irregularities prominent with lower Q curves, and larger positive than negative jitter. The former can be explained by taking jitter-fine structure (previewed in Section 1.7 and further studied in Figure 7.5) into account—the periods of irregularities coincide with those of fine structures.

We can understand the latter by checking Figure 7.2. We find that all quasi-static jitter waveforms in the figure have rise times slower than their fall times. We can explain this phenomenon with the flywheel effect of narrow-band filters for clock extraction (Problem 7.2). Note that the leading and trailing edges of quasistatic jitter waveforms correspond to negative and positive peaks of extra-alignment jitter (see Figures 1.18 and 7.5).

Careful readers will see that quasistatic jitter waveforms for the 1/8–8/8 pattern have slower rise times but faster fall times, compared to pattern 1/2–8/8 (Problem 7.3). This seems to explain the discrepancies in the location of positive and negative peaks in the curves of Figure 7.3(b) (Problem 7.4).

7.2.2 Jitter Accumulation in a Chain

Definitions of quasistatic jitter $\theta(t)$ for the study of jitter accumulation are listed in Table 7.2. Note the difference with $\theta(t)$ for theoretical study in Chapter 6, also listed in the table. Peak-to-peak jitter for the 1/2–8/8 pattern is designed to be almost equal to that used in theoretical study, because we plan to mainly use this pattern in the simulation.

Note that rise and fall times are defined as zero in the theoretical study, an element that may lead to some errors in theoretical predictions and will be discussed below. A shorter pattern period is used in the simulation due to the limit in computation capacity.

Results for the processing delay of $M = 0$ are shown in Figure 7.4. By comparing these results with those in Figure 6.9, we find that theoretical predictions in Figure 6.9 suffer from a higher degree of band limitations than shown in Figure 7.4. This is because high-frequency fine structures tend to be lost when we study jitter accumulation with Chapman's model. As a consequence, we can see that Figure 6.9(a) provides a better prediction of Figure 7.4(a) than does Figure 6.9(a'). Also, we can see that Figure 6.9(b) is a better prediction of Figure 7.4(b).

Results for processing delay M that maximize extra-accumulation are shown in Figures 7.5, 7.6 and 7.7, for Q values of 100, 500 and 2000, respectively. The theoretical predictions in Figures 6.10(a) and (b) exhibit good agreement with the results in Figures 7.5(a) and (b), respectively.

The use of Chapman's model does not entail a significant problem as far as jitter-fine structures shown in Figure 7.5(c) are concerned. We see there that errors in theoretical predictions are not so large, since these fine structures do not accumulate along the repeater chain. Note in Figure 7.5(d) that the fine structures of

Table 7.2

Definitions of QuasiStatic Jitter $\theta(t)$ for the Study of Jitter Accumulation

		Simulation in Chap. 7*		Theoretical
		1/2-8/8 pattern	1/8-8/8 pattern	Study in Chap. 6
Peak to Peak	Q= 100	11.5	6.3	
Jitter	Q= 500	11.1	5.7	10
(Degree)	Q=2000	9.9	4.1	
Rise Time	Q= 100			
	Q= 500			0
	Q=2000	see Fig. 7.2		
Fall Time	Q= 100			
	Q= 500			0
	Q=2000			
Pattern				4096T and
		4096T	4096T	
Period				132768T

* Type B jitter caused by echo #2 in Fig. 5.6 with ε=0.05.

alignment jitter are doubled at the second repeater ($N = 2$), then kept constant for N values larger than 2 (Problem 7.5).

Although Figure 6.11(a') is not necessarily a good prediction of Figure 7.6(a), it is interesting to note that alignment jitter in Figure 7.6(b) is foreseen with relatively high accuracy by Figure 6.11(b'). We may expect that, should we apply $T_p = 8192T$ for predictions in Figure 6.12, better predictions for Figure 7.7 would result.

The results of computer simulation for jitter accumulation in a chain of 100 repeaters are summarized in Figure 7.8. We can compare those results with theoretical predictions shown in Figure 6.13. As we have already discussed, we must consider the fact that the theoretical prediction is based on Chapman's model. That is, Figures 6.13 (a) and (b) suffer from a lower degree of band limitation and therefore can provide better predictions than Figures 6.13 (a') and (b').

Let us first compare Figure 6.13 (a) and Figure 7.8 (a), based on the above understanding. We can see that theory and simulation agree insofar as the number

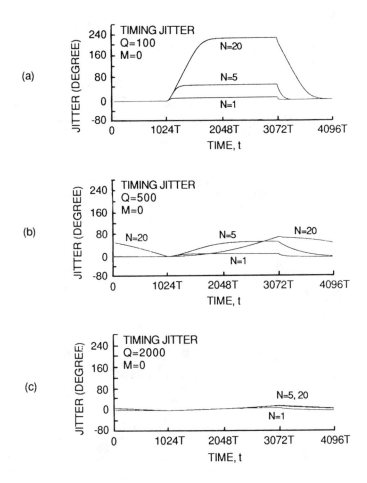

Figure 7.4 Timing jitter accumulation: (a) $Q = 100$, (b) $Q = 500$, and (c) $Q = 2000$.

of repeaters (N) in chain is relatively small. However, systems with smaller Q values become unstable when N exceeds 30.

Next, let us compare Figures 6.13(b) and 7.8(b) to investigate the cause of such unstableness. Note that the Chapman's model assumes that jitter adds linearly from repeater to repeater (see Section 6.1). We can use this assumption in most practical systems, where jitter changes very slowly, or where high-frequency components of jitter is sufficiently small. We cannot apply this assumption when the amount of alignment jitter becomes relatively large. We can see in Figure 7.8 that the behavior of alignment-jitter accumulation starts to change when alignment jitter exceeds around 50°, and that systems become unstable when the amount of jitter

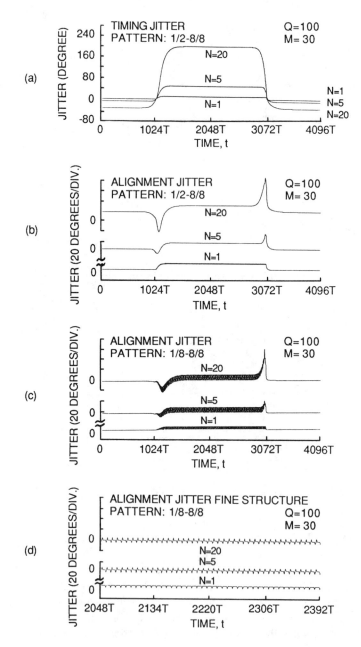

Figure 7.5 Worst-case extra-jitter accumulation for $Q = 100$ ($M = M_w = 30$): (a) timing jitter for pattern 1/2–8/8; (b) alignment jitter for pattern 1/2–8/8; (c) alignment jitter for pattern 1/8–8/8; and (d) fine structures of jitter for pattern 1/8–8/8.

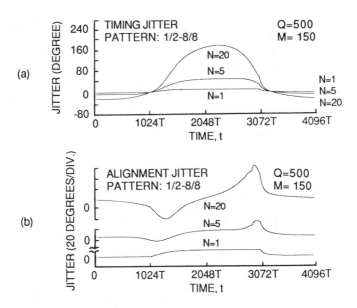

Figure 7.6 Worst-case extra-accumulation for $Q = 500$ and pattern $1/2$–$8/8$ ($M = M_w = 150$); (a) timing jitter, and (b) alignment jitter.

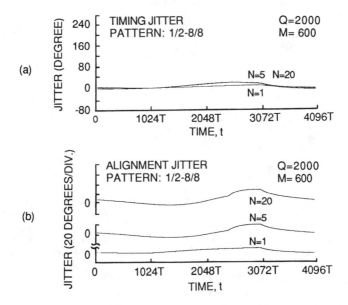

Figure 7.7 Worst-case extra-accumulation for $Q = 2000$ and pattern $1/2$–$8/8$ ($M = M_w = 600$).

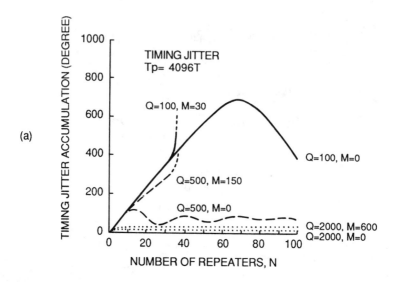

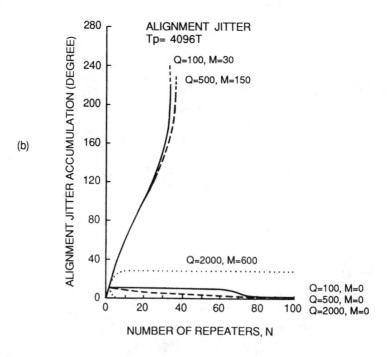

Figure 7.8 Summary of computer simulation for jitter accumulation in a chain of 100 repeaters; (a) timing jitter, and (b) alignment jitter.

exceeds approximately 180°. It should be evident that we cannot apply Chapman's model in such an environment. As a matter of practicality, such environments are avoided in practical designs.

7.2.3 Alleviating Extra-Accumulation

Let us check, by simulation, the second and third methods for alleviating extra-accumulation discussed in Section 6.2.3. The results of computer simulation for the second method are shown in Figure 7.9. Theoretical predictions in Figure 6.14 show good agreement when T_p is relatively large. Poor predictions for smaller T_p values may be attributed to the use of quasistatic jitter where the rise and fall times are assumed to be zero (see Table 7.2): slower rise and fall times can alleviate alignment jitter for smaller T_p values.

The results of computer simulation for the third method are shown in Figure 7.10. They are in good agreement with the theoretical prediction in Figure 6.15. This is because the amount of alignment jitter is relatively small, and the applicability of Chapman's model is considerably good.

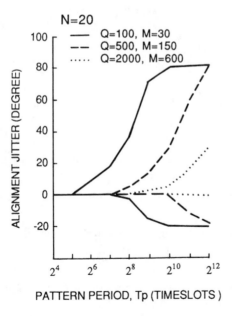

Figure 7.9 Dependence of extra-alignment jitter accumulation on the period of quasistatic pattern 1/2–8/8.

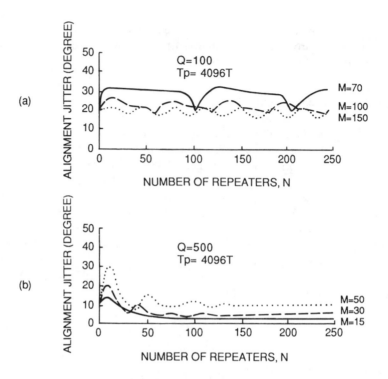

Figure 7.10 Alleviating extra-alignment jitter accumulation by deviating Q/M ratios from 3: (a) deviating Q/M ratios above 3 for $Q = 100$, and (b) deviating Q/M ratios below 3 for $Q = 500$.

PROBLEMS

Problem 7.1 Explain why a quasistatic pattern of $1/2$–$8/8$ contributes the largest amount of jitter, compared to other patterns (Figure 7.2).

Problem 7.2 Explain why quasistatic waveforms in Figure 7.2 have slower rise times than fall times.

Problem 7.3 Explain why quasistatic jitter waveforms for the pattern $1/8$–$8/8$ in Figure 7.2 have slower rise times, but faster fall times compared to the $1/2$–$8/8$ pattern.

Problem 7.4 Speculate on why the locations of negative and positive peaks of curves in Figure 7.3(b) do not coincide.

Problem 7.5 Speculate on why the fine structures of alignment jitter are doubled at the second repeater and then kept constant in Figure 7.5(d).

Chapter 8

Transmission System Design

We study the design of amplitude- and time-crosshair budgets based on worst-case, rms, and hybrid criteria. We use practical examples to understand the usefulness and applicability of the hybrid criterion.

Next, we discuss the economy of transmission networks from the standpoint of transmission media selection. We can see that the cost of local networks is the dominant component of total system cost. We study the efficiency of applying metallic cables in reducing the cost of local networks. Based on the fact that optical fibers have proved extremely efficient in trunk line applications, we also investigate the longer-term applicability of fiber optics to local systems.

Finally, we summarize how to apply the theory and technologies studied in this book with some practical examples. We consider a metallic cable system to see how we can modify the theory of variable equalizers to suit a practical system. We also study a fiber optic transmission system to understand how to select and design line coding for digital transmission.

8.1 TRANSMISSION SYSTEM DESIGN

We learned in Section 1.1 that the design of crosshair budget based on worst-case criterion can lead to excessive requirements in digital transmission design. We take the results of Chapters 6 and 7 into consideration in Section 8.1.1 to see how we can modify the design of crosshair budget.

Also discussed in Section 1.1 is the fact that the cost of subscriber loops, or local networks, dominates total system cost, and that the selection of transmission media is important in reducing the cost of such local systems. We study simple cost analyses to see the influence of transmission media selection on system economy.

8.1.1 Design Criteria

We studied the worst-case design of crosshair budget in Section 1.1. We now understand that it is difficult to assign sufficient budgets for degradation in time crosshair due to the extra-accumulation of alignment jitter. Budget design based on the rms criterion could alleviate this problem. However, the rms criterion is usually too optimistic to guarantee a reliable design.

As a compromise, let us study a hybrid (worst-case combined with rms) design. Typical examples of worst-case and hybrid designs are compared in Table 8.1. The

Table 8.1
Example of Crosshair Budget Design

Design Criterion	Worst-Case	Hybrid
Amplitude Crosshair	(60 %)	(45 + 8.5 %)
Noise	25 %	20 + 5 %
Intersymbol Interference	25 %	20 + 5 %
Threshold Displacement	10 %	5 + 5 %
Time Crosshair	(13 %)	(15 + 4 %)
Clock Phase Displacement	±45°	±(32 + 7.5)°
Alignment Jitter	±15°	± 32 + 0°
Margin	27 %	30.5 %
Total	100 %	90.5 + 9.5 %

Table 8.2
Example of Time Crosshair Budget Design

Design Criterion	Worst-Case	Hybrid
Clock Phase Displacement	(± 45°	± [32 + 7.5°])
Initial Phase Errors	± 15°	± [10 + 5°]
Temperature Fluctuations	± 20°	± [15 + 5°]
Aging	± 10°	± [7 + 3°]
Alignment Jitter	(± 15°	± [32 + 0°])
Type A	± 5°	± [5 + 0°]
Type B (N = 256)	± 10°	± [27 + 0°]
Total	± 60°	± [64 + 7.5°]
(Amplitude crosshair equivalent)	(13 %)	(15 + 4%)

first and second terms in the hybrid design budgets are worst-case and rms budgets, respectively. Note that the budget for alignment jitter can be increased considerably without loosing margin.

The details of time crosshair budgets are shown in Table 8.2. It was customary to assign ±10° for alignment jitter. However, as we learned in Section 6.2.1, we must take not only type B, but also type A jitter into consideration. A larger budget is especially required with type B (or type B′) jitter. We can see that the hybrid criterion is again useful in alleviating such a problem in time crosshair design. We can use the curve in Figure 8.1 for converting time crosshair degradation into equivalent degradation in amplitude crosshair (Problem 8.1).

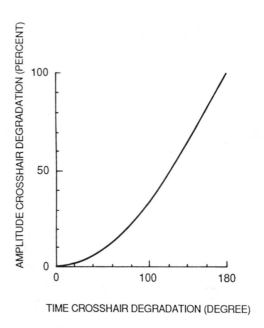

Figure 8.1 Converting time crosshair degradations into equivalent amplitude crosshair degradations.

8.1.2   CHOICE OF TRANSMISSION MEDIA

We learned in Section 1.1 that the importance of applying metallic cable transmission in local systems is increasing, at least in the short term. This is simply because metallic cable systems are still more economical than fiber optic systems in local network applications, even at relatively high transmission rates. Let us study this secret by using some examples.

Fiber optic transmission and metallic cable transmission are compared in Figure 8.2 in terms of transmission capacity *versus* repeater spacing. It is evident from Figures 8.2(a) and (c) that fiber optic transmission is far superior to metallic cable transmission with regard to repeater spacing. However, in local systems, a repeater spacing of more than several kilometers may be of no use. Moreover, the use of a long-wavelength laser diode and single-mode fiber for attaining large capacity can adversely affect economy in local networks where the multiplexing rate is relatively low.

Let us assume that the repeater spacing is 100 m and compare the cost of transmission line for a capacity of 150 Mb/s. The results are shown in Figure 8.3 and were calculated under the cost assumptions (near-term target costs in quantity) listed in Table 8.3 [1]. As is apparent, metallic cable transmission is much more

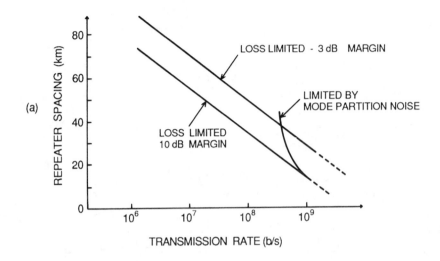

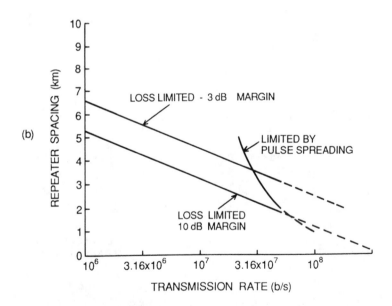

Figure 8.2 Comparison of fiber optic and metallic cable transmissions; (a) long-wavelength single-mode fiber transmission; (b) short-wavelength multimode fiber transmission; *Source*: Personick S.D., Fiber Optics, Technology and Applications, New York: Plenum Press, 1985, pp. 136–144. Reprinted with permission. Copyright© 1985 Plenum Press; and (c) Metallic cable transmission.

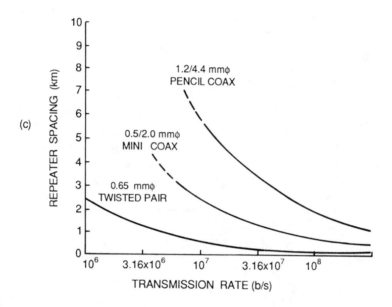

Figure 8.2 Continued.

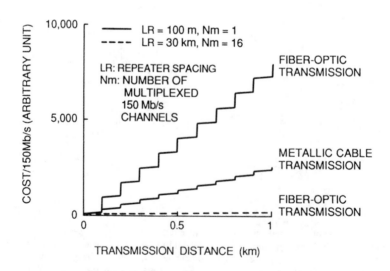

Figure 8.3 Cost of digital transmission lines.

Table 8.3
Cost Assumptions (in Dollars)

Electrical	TX-RX	(150 Mb/s)	100 $\sqrt{3}$
Optical	TX-RX	(150 Mb/s)	400 $\sqrt{3}$
Optical	TX-RX	(2.4 Gb/s)	$400\sqrt{48}$
Metallic	Cable	(1 m)	0.75
Optical	Fiber	(1 m)	1

economical than fiber optic transmission. As shown in the figure by a dotted line, fiber optic transmission can become economical for relatively high multiplexing rate $N_m = 16$.

One way to reduce the cost of fiber optic transmission is to use a multimode fiber, a light emitting diode (LED), and a *pin* photodetector at the sacrifice of repeater spacing. Such a system is applicable to local networks, as seen from Figure 8.2(b). While they still may be more expensive than metallic cable systems (Problem 8.2), fiber optic systems can take advantage of larger immunity to interference from other systems.

One of the major problems in designing metallic cable transmission is the equalization of cable frequency characteristics, which will be discussed in Section 8.2. On the other hand, a multimode fiber optic system using an LED suffers from band limitation due to material dispersion [2]. We study an application of line coding in Section 8.3 for alleviating such a problem. We learn, through these examples, how the theory and technologies studied in this book can be modified to suit practical applications.

8.2 METALLIC CABLE SYSTEM

Metallic cable transmission suffers from dramatic changes in cable frequency responses, as shown in Figure 8.4, due to the change of parameters such as repeater

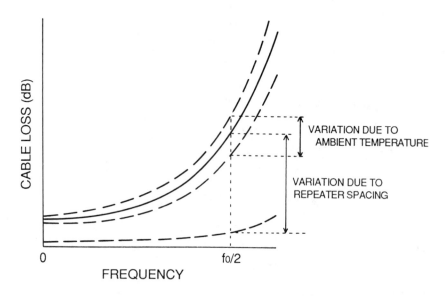

Figure 8.4 Frequency responses of a metallic cable.

spacing and ambient temperature. Loss variations at the equivalent transmission frequency $f_0/2$ can exceed 30dB. ALBO technology [3] has been developed for compensations of such loss variations in metallic cables. As we learned in Chapter 5, equalization errors must be made as small as possible from the standpoint of reducing timing jitter. For realizing sufficiently small equalization error over the ALBO networks' wide variable range, the cascading of two variable equalizers has been considered in conventional PCM repeaters. For example, cascading of two Tarbox-type variable equalizers [3] has been reported [4]. A Bode-type variable equalizer [5] is potentially suitable for realizing ALBO with a single stage. However, realization of one-port dual-shaping networks has been the bottleneck for attaining a wide variable range. Equalizer design has also been quite cumbersome.

We study a simple, wide-range, inductorless variable equalizer in this section, as an application of theory described in Section 3.2. We also study the influence of equalization error on jitter generation.

As we learned in Section 3.2, an ideal variable transfer function is given by

$$v_0(f,u) = y_0(f)^u \tag{8.1}$$

where

$$y_0(f) = \text{frequency response of a cable}$$

$$f = \text{frequency}$$

$$u = \text{a variable corresponding to a parameter such as cable length}$$

The Bode-type variable transfer function (bilinear variable transfer function)

$$v(f,x) = [x + y_0(f)]/[y_0(f)x + 1], \quad (0 \le x \le \infty) \tag{8.2}$$

approximates the desired variable transfer function (8.1), where variable x can be related to u by $u = (1 - x)/(1 + x)$. Let us assume a variable range of ± 15 dB at equivalent transmission frequency fe $(=f_0/2)$, where f_0 is pulse repetition frequency. As we studied in Section 3.2, a straightforward realization of (8.2) requires an amplifier with a gain of 20 dB for the implementation of variable x. The saturation of the amplifier with a large input signal could become a problem with such a realization. An x range modification technique [6] can be applied to alleviate this problem. That is, an x range of $0 \le x \le \infty$ can be converted to $0 \le x \le 1$ by applying the bilinear conversion $x \rightarrow \{(1 - x_m)/x_m\}\{x/(1 - x)\}$, $(0 \le x \le 1)$, to (8.2). We obtain

$$v(f,x) = (x_m^{-1} - 1 - y_0)[x + y_0(x_m^{-1} - 1 - y_0)^{-1}]/[(x_m^{-1}y_0 - y_0 - 1)x + 1] \tag{8.3}$$

with $u = (1 - x)/[1 - (2 - x_m^{-1})x]$

Equation (8.3) can be realized as shown in Figure 8.5(a) except for the multiplying constant [7]. That is, the transfer function of this realization is described as

$$v_m(f,x) = v(f,x)/(x_m^{-1} - 1 - y_0) \tag{8.4}$$

There are three advantages to this new equalizer. It can realize a wide variable-element adjusting range, it has relatively simple structure and requires no inductors, and its design is quite straightforward, as we will see below.

The term $-y_0$ in the transfer function of the feed-forward path in Figure 8.5(a) can be neglected by choosing x_m sufficiently small. In this case, the equalizer in (a) can be realized as shown in (b). The design is straightforward because $y_0(f)$ corresponds to the frequency response of a cable for transmission. We will study below the fact that the realization requires no inductors for the ALBO application. The use of two-port shaping networks (as opposed to conventional variable equalizers that use one-port shaping networks) helps the realization of a wide variable range. Note that the adjusting range of variable element x is very small, since only $0 \le x \le 1$ is sufficient, as opposed to requirement $0 \le x \le \infty$ for conventional variable equalizers.

Figure 8.6 illustrates the variable characteristics and maximum errors for an ALBO with a ± 15 dB variable range. Note the difference in solid and dotted lines in the error curve. We can see that the error increments due to approximating the circuit in Figure 8.5(a) with that in (b) are negligibly small.

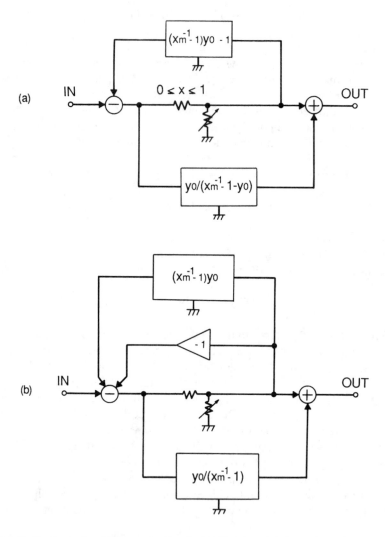

Figure 8.5 Realizations of variable transfer function (8.3); (a) straightforward realization, and (b) approximate realization. *Source*: Takasaki Y., "Simple Inductorless Automatic Line Equalizer for PCM Transmission Using New Variable Transfer Function," *IEEE Trans. Commun.*, Vol. COM-26, No. 5, May 1978, pp. 675–678. Reprinted with permission. Copyright© 1978 IEEE.

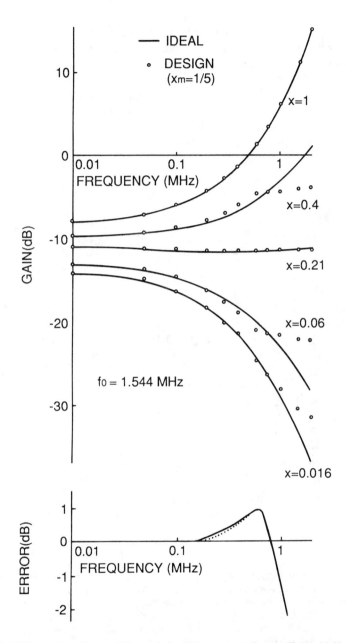

Figure 8.6 Variable transfer characteristics and simulation errors. *Source*: Takasaki Y., "Simple Inductorless Automatic Line Equalizer for PCM Transmission Using New Variable Transfer Function," *IEEE Trans. Commun.*, Vol. COM-26, No. 5, May 1978, pp. 675–678. Reprinted with permission. Copyright© 1978 IEEE.

Now, let us study the efficiencies of this new variable equalizer applied to PCM transmission with AMI line coding. ISI due to equalization errors, phases of waveform functions, and static timing jitter for several AMI patterns are plotted in Figure 8.7(a), (b) and (c), respectively. For calculating the plots in the figures, a fifth-order Bessel shaping (with a 5 dB loss at the equivalent transmission frequency $f_0/2$) and 10% nonlinear equalization are assumed (see the inset of Figure 8.7(a)). We can see that ISI and jitter due to equalization error are acceptable from the standpoint of crosshair budget design shown in Table 8.1 (Problem 8.3).

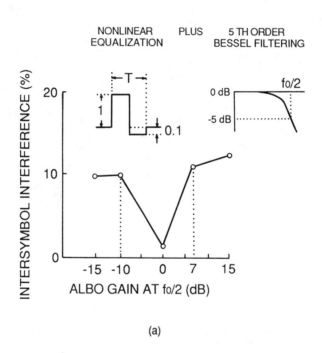

(a)

Figure 8.7 Efficiencies of variable equalizer in Figure 8.5(b): (a) intersymbol interference *versus* automatic line build-out gain; (b) phases of waveform functions; and (c) pattern-dependent jitter for AMI sequences. *Source*: Takasaki Y., "Simple Inductorless Automatic Line Equalizer for PCM Transmission Using New Variable Transfer Function," *IEEE Trans. Commun.*, Vol. COM-26, No. 5, May 1978, pp. 675–678. Reprinted with permission. Copyright© 1978 IEEE.

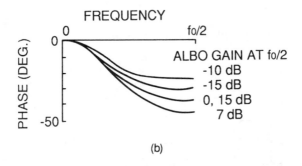

(b)

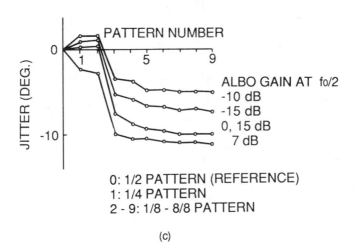

0: 1/2 PATTERN (REFERENCE)
1: 1/4 PATTERN
2 - 9: 1/8 - 8/8 PATTERN

(c)

Figure 8.7 Continued.

An embodiment of the signal flow graph in Figure 8.5(b), designed to realize the characteristic specified in Figure 8.6, is illustrated in Figure 8.8. It is seen that very simple RC networks (enclosed by dotted lines) are used for the approximate realization of cable response $y_0(f)$ (see Figures 8.4 and 5). Frequency responses of the equalizer in the figure are plotted in Figure 8.9. In a practical application for a 1.544 Mb/s PCM repeater, a fixed equalizer is added as shown in Figure 8.10(a). Note that the fixed equalizer can be included in the variable equalizer to attain network simplification as shown in Figure 8.10(b).

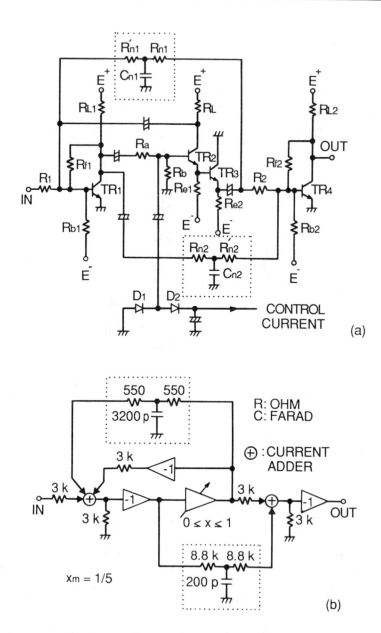

Figure 8.8 Embodiment of the signal flow graph in Figure 8.5(b); (a) inductorless realization, and (b) equivalent expression of the circuit in (a). *Source*: Takasaki Y., "Simple Inductorless Automatic Line Equalizer for PCM Transmission Using New Variable Transfer Function," *IEEE Trans. Commun.*, Vol. COM-26, No. 5, May 1978, pp. 675–678. Reprinted with permission. Copyright© 1978 IEEE.

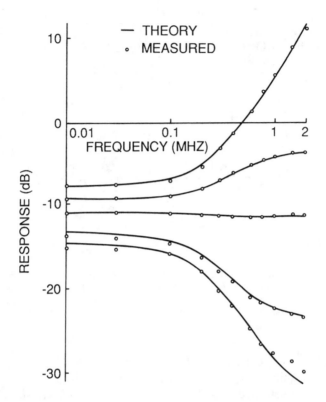

Figure 8.9 Frequency responses of the variable equalizer in Figure 8.8(a). *Source*: Takasaki Y., "Simple Inductorless Automatic Line Equalizer for PCM Transmission Using New Variable Transfer Function," *IEEE Trans. Commun.*, Vol. COM-26, No. 5, May 1978, pp. 675–678. Reprinted with permission. Copyright© 1978 IEEE.

8.3 OPTICAL FIBER SYSTEM

Fiber optic transmission has proven very attractive for trunk line applications. As we studied in Section 8.1, it is still too expensive to be applied in local systems, where repeater spacing is relatively short. Drastic reduction of optical device and component costs, as well as installation costs, will be necessary before optical fiber systems become more economical than copper cable systems. It is interesting to note that an effort to reduce cost always leads to the problem of bandwidth limitation.

In this section we will study a digital fiber optic transmission system that was designed for a local area network. A method for the efficient use of bandwidth will be described. The cost of optical fiber is not dominant, since transmission distance is rather short. The use of a multimode fiber is preferable from the standpoint of

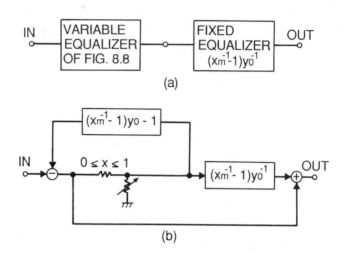

Figure 8.10 Composite equalizer; (a) variable plus fixed equalizer, and (b) simplified realization. *Source* Takasaki Y., "Simple Inductorless Automatic Line Equalizer for PCM Transmission Using New Variable Transfer Function," *IEEE Trans. Commun.*, Vol. COM-26, No. 5, May 1978, pp. 675–678. Reprinted with permission. Copyright© 1978 IEEE.

connector and splicing cost. Also, the selection of optical source and detector is important for designing an economical system.

The least sophisticated digital fiber optic transmission system would employ an LED source, a multimode fiber, and a *pin* photodiode (PD) detector. A system using the short wavelength (0.8–0.9 μm) region would draw on the more mature GaAlAs source and S_i detector technology. To achieve improved performance (higher data rates and longer transmission distances) we can also use a GaAlAs laser. The use of laser diodes (LDs) developed for compact disc application may contribute to the design of economical systems.

Typical examples of the average receiving power *versus* transmission speeds required are plotted in Figure 8.11. We can see that the PD has 10 to 15 dB lower sensitivity than the avalanche PD. Optical power of around 0 dBm and −15 dBm can be coupled to an optical fiber when we use the LD and the light-emitting diode (LED), respectively. Repeater spacing can be calculated if the loss of optical fiber is given.

Note, however, that we must take material dispersion and mode dispersion into consideration when we calculate repeater spacing [2]. Boundaries for the limitation of repeater spacing due to fiber loss and bandwidth are drawn in Figure 8.11 by broken lines. The boundary on the left is attributed to material dispersion caused by the broad spectral width of the LED. The one on the right shows the influence of mode dispersion by a multimode fiber where an LD is used as an optical source.

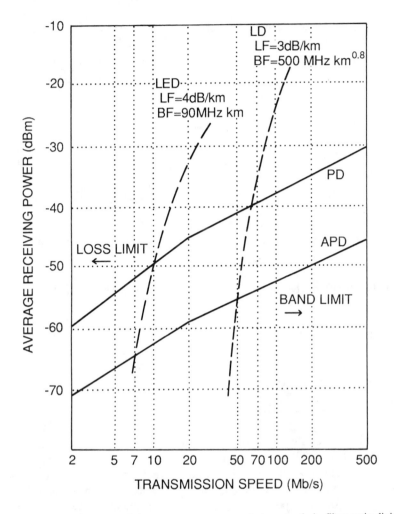

Figure 8.11 Required average receiving power *versus* transmission speeds in fiber optic digital transmission. After Y. Takasaki and Y. Takahashi [8].

The use of short-wavelength optical sources is assumed in both cases. Figure 8.11 also shows that using an LD is preferable from the standpoint of realizing a larger transmission capacity and longer repeater spacing. However, the laser requires a more complicated driver circuit and is more sensitive to electrical abuse and temperature. Systems employing laser sources are also susceptible to modal noise and mode partition noise. Therefore, the use of an LED is almost mandatory from the standpoint of realizing economical and reliable systems.

Major specifications of a sample system we will design in this section are shown in Table 8.4 [8]. Its major design parameters are shown in Table 8.5. Note that the average optical power of -19 dBm has to be guaranteed at the highest temperature of $50°$ C.

Table 8.4
Major Specifications

Optical Source	LED(0.83 μm)
Detector	PIN Photodiode
Information Rate	32.064 Mb/s
Error Rate	$< 10^{-15}$
Repeater Spacing	0 ~ 3 km

Several line coding plans were compared, taking the specifications in Table 8.4 into consideration. Line coding formats such as *mBnB* are not desirable, since they require the use of a scrambler for reliable synchronization of coded blocks, and a scrambler may fail to satisfy the condition of bit sequence independence (BSI), or tolerance to any particular pulse stream pattern. Also, implementation could get costly. The partitioned-block 8B/10B code [9] may be applicable, provided an LSI coder is available. Such redundant formats as coded and differential mark inversion, or CMI and DMI, and Manchester code are very simple but require a bandwidth twice as large, (i.e., 64 MHz) and therefore cannot satisfy the specification on repeater spacing.

We can solve this problem by using the DMI-duobinary scheme in Figure 8.12, which we studied in Section 2.5.4. At the transmitter, the original information signal (a) is converted into a DMI pulse stream (b). It can be shown that the influence of band limitation by the transmission medium can be alleviated by applying duobinary filtering at the receiver. As seen in the figure, received pulse stream (c) is delayed by half a timeslot ($T/2$) and delayed pulse stream (d) is added to pulse stream (c)

Table 8.5
Major Design Parameters

Optical Fiber Loss	4 dB/km	including aging and splicing
LED Spectral Width	$\Delta\lambda$ = 50 nm	
Material Dispersion	4500 MHz·nm·km = 90 MHz·km(for $\Delta\lambda$ = 50 nm)	
Fiber Input Power	-19 dBm	(average at 50°C)
Ambient Temperature	0 ~ 50°C	

o result in a three-level pulse stream (e). (Waveforms (c), (d) and (e) are used without band limitation for the convenience of explanation.) Although duobinary filter $D(f)$ has an infinite bandwidth, composite filter $H(f) = D(f)R(f)$ is band imited within f_0, as seen from Figure 8.13, where $R(f)$ represents a raised cosine filter.

We can see that a transmitted two-level DMI signal is received as a three-level AMI signal. That is, horizontal (time axis) redundancies are converted into vertical amplitude axis) redundancies. One problem that remains to be solved is the loss of iming information with a long succession of zeros. A solution by means of a ZS echnique is explained in Figure 8.14 [10]. A succession of eight zeros of the DMI ormat, "0101010101010101" (waveform (b)), which will cause eight successive zeros after duobinary filtering, is substituted by "0010010011011011" (waveform b')). It can be shown that duobinary filtering of waveform (b') results in waveform e'). This waveform, after raised cosine filtering, results in waveform (f'). Note that a raised cosine filtering with a roll-off frequency of $f_0/2$ is applied, instead of an deal low-pass filtering with a cutoff frequency of f_0 (f_0 = pulse repetition fre-quency). This leads to somewhat different waveforms from those encountered in the

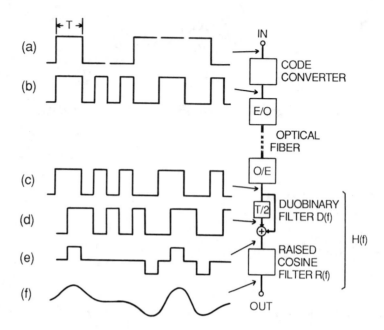

Figure 8.12 Principles of DMI-duobinary transmission: (a) pulse sequence with NRZ format; (b) and (c) pulse sequence with DMI format; (d) pulse sequence (b) or (c) delayed by $T/2$; (e) pulse sequence (c) after duobinary filtering; and (f) pulse sequence (e) after raised cosine filtering. *Source*: Takasaki Y., and Y. Takahashi, "DMI/Duobinary Transmission Experiment in a Chain of Fiber-Optic Repeaters," *Int. Conf. Commun. '83*, June 19–22, 1983, Boston MA, pp. 100–104. Reprinted with permission. Copyright© 1983, IEEE.

conventional duobinary scheme we studied in Chapter 2. The pulses indicated by arrows are considered undesirable, because they include timing information, the polarity of which is inverted, compared with the case for a normal pulse (Problem 8.4). However, the substitution pattern is designed so the number of normal pulses (those without arrows) exceeds that of undesirable pulses.

A final problem is how to eliminate the substituted waveform at the receiver. This procedure can be explained with the help of waveform (g'). Waveform (g') is obtained by regenerating waveform (f'). The dotted lines in waveform (f') are positive and negative thresholds for pulse decision. It is seen that waveform (f') comes very close to these thresholds at retiming points just before, and just after, undesirable pulses. Thus, pulse regeneration becomes quite sensitive to noise conditions at these decision points. The broken lines in waveform (g') indicate undefined pulses:

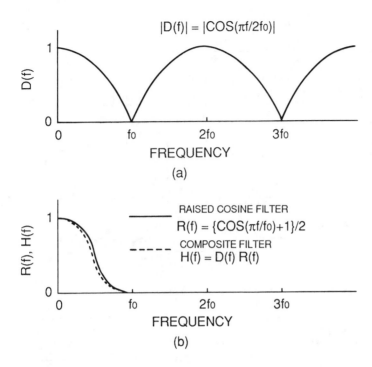

Figure 8.13 Composite duobinary-raised cosine filtering; (a) duobinary filter, and (b) raised cosine filter and composite filter.

the regeneration of pulse depends completely on noise polarity. The part of zero substitution in the regenerated pulse sequence can be expressed as

$$-1, \quad 0(-), \quad 0(-), \quad -1, \quad 1, \quad 0(+), \quad 0(+), \quad 1$$

where $0(+)$ and $0(-)$ indicate undefined pulses, ($+$ or 0), and ($-$ or 0), respectively.

Detecting pattern $[1, 0(+), 0(+), 1]$ or $[-1, 0(-), 0(-), -1]$ can be used to provide information for eliminating ZS. Note that the number of consecutive pulses we must check for detecting ZS is only four, not eight. In studying error propagation we can neglect the probability that $0(+)$ or $0(-)$ is regenerated as -1 or 1, respectively (Problem 8.5). Then, a typical worst-case error propagation occurs when "1, 1, 1, 1" (a part of zero substitution), is regenerated as "0, 1, 1, 1."

In this case "1, 1, 1, 1" (which should be decoded as "0, 0, 0, 0") becomes "0, 1, 1, 1." This way, the error propagation due to ZS can be limited to within three timeslots at the most.

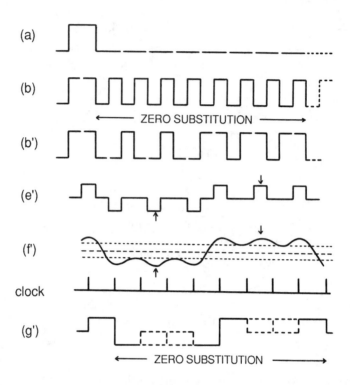

Figure 8.14 Zero-substitution technique for DMI-duobinary transmission: After Y. Takasaki [10]. (a) NRZ pulse sequence with eight-zero succession; (b) DMI pulse sequence with eight-zero succession; (b') DMI pulse sequence with eight-zero substitution (8ZS); (e') duobinary filtered version of sequence (b'); (f') raised cosine filtered version of sequence (e'); and (g') sequence (f') after regeneration.

Now, we can allocate the loss budgets for designing a DMI-duobinary system, as shown in Table 8.6. A receiver sensitivity of −39 dBm, one that includes not only impairments by receiver noise (see Figure 8.11), but several types of degradations in amplitude, as well as time crosshairs (mentioned in Section 8.1.1) is required. A schematic diagram of an optical regenerator for DMI-duobinary transmission is shown in Figure 8.15(a). Composite (duobinary-raised cosine) filtering function $H(f)$ can be realized by a simple RLC network (enclosed by dotted lines). A timing tank Q of around 70 was found to suffice for stable clock recovery, since the longest zero succession is kept to less than 7, owing to the 8ZS mentioned above.

The decoder is used for detecting and eliminating ZS patterns. A sample decoder circuit is shown in Figure 8.15(c). The decoder uses properties of positive (P)

Table 8.6
Loss Budgets

Transmitted Power	-19 dBm	
Fiber Loss	12 dB	(3 km)
Band Limitation	5 dB	(32 MHz·3 km)
Margin	3 dB	
Receiver Sensitivity	-39 dBm	

and negative (N) pulses during the period of ZS shown in Table 8.7 for decoding a pulse stream with ZS pulses. As seen in the table, ZS can be detected by checking four consecutive timeslots: it is determined to be a ZS pattern when the first and fourth timeslots of $P(N)$ pulses are marks, and the second and third timeslots of $N(P)$ pulses are spaces. A waveform after composite filtering $H(f)$ is shown in Figure 8.16 for a repetitive pattern of "1, 1, 8 ZS."

The coder is used for ZS and DMI coding. A sample coder circuit is shown in Figure 8.15(b). When eight successive zeros are detected, the shift register is set to generate a "10011001" pattern. This pattern is used to inhibit clock pulse ϕ_1, and obtain pulse train p_1, in Figure 8.17. Similarly, pulse train p_2 can be obtained by inhibiting clock pulse ϕ_2 (Problem 8.6). Composite pulse train p_3 is applied to the trigger-type flip-flop to produce ZS waveform (b') in Figure 8.14.

Accumulation of static-pattern jitter in an experimental system is shown in Figure 8.18. The amount of jitter at the first regenerator is around 2 ns, but this is thought to include type A, or nonsystematic, components that do not accumulate systematically. It can be seen in that figure that type B systematic jitter is around 1 ns per repeater (i.e., ±6° for 32 Mb/s transmission).

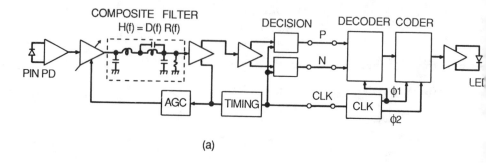

(a)

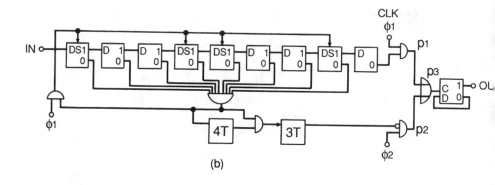

(b)

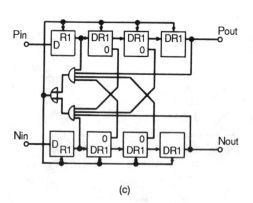

(c)

Figure 8.15 Schematic diagram of DMI-duobinary regenerator: (a) optical regenerator; (b) DMI coder with zero-substitution function; and (c) AMI decoder with zero-substitution eliminator. *Source*: Takasaki Y., and Y. Takahashi, "DMI/Duobinary Transmission Experiment in a Chain of Fiber-Optic Repeaters," *Int. Conf. Commun. '83*, June 19–22, 1983, Boston MA, pp. 100–104. Reprinted with permission. Copyright© 1983, IEEE.

Table 8.7
P and N Pulses at Decoded ZS

Decoded ZS	-1,	0(-),	0(-),	-1,	1,	0(+),	0(+),	1
P Pulses	0,	0,	0,	0,	1,	0/1,	0/1,	1
N Pulses	1,	0/1,	0/1,	1,	0,	0,	0,	0

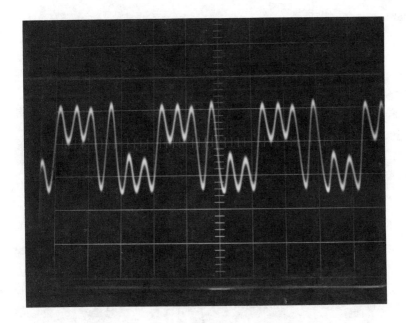

Figure 8.16 Waveform with 8ZS after composite filtering.

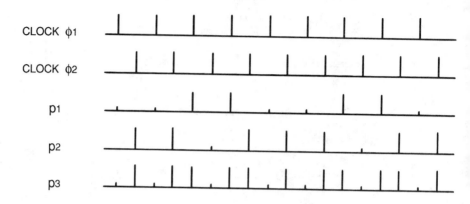

CLOCK φ1

CLOCK φ2

p1

p2

p3

Figure 8.17 Time chart for DMI coder with zero substitution function in Figure 8.15(b).

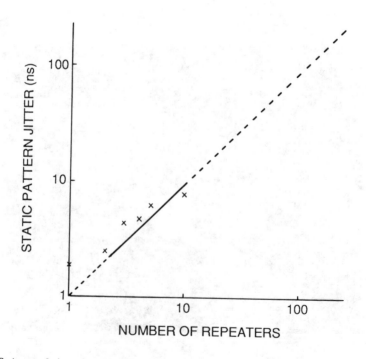

Figure 8.18 Accumulation of static-pattern jitter in an experimental system.

PROBLEMS

Problem 8.1 Explain how we can get the curve in Figure 8.1, and how we can use it.

Problem 8.2 Compare an LED multimode fiber system with a metallic cable system. Assume the cost of the optical transmitter-receiver pair in Table 8.3 can be halved.

Problem 8.3 Explain why maximum jitter in Figure 8.7(c) is less than 11°, while that in Figure 8.7(b) is around 45°.

Problem 8.4 Explain the influence of the pulses indicated by arrows in Figure 8.14.

Problem 8.5 Explain why we can neglect the probability that $0(+)$ or $0(-)$ is regenerated as -1 or 1, respectively, in Figure 8.14.

Problem 8.6 What is the rule for inhibiting clock pulse ϕ_2 to obtain pulse train p_2 in Figure 8.17?

REFERENCES

1. Takasaki Y., Upgrading Strategies for B-ISDN Subscriber Loops, *IEEE J. Lightwave Tech.*, Vol. 7, No. 11, November 1989, pp. 1778–1787.
2. Personick S.D., Fiber Optics, Technology and Applications, New York: Plenum Press, 1985, p. 19.
3. Tarbox, R.A., A Regenerative Repeater Utilizing Hybrid Integrated Circuit Technology, *Proc. Int. Conf. Commun.*, June 1969, Paper #69C.
4. Anuff A., *et al.*, A New 3.152 MB/s Digital Repeater, *Proc. Int. Commun.*, June 1975, pp. 39-10–13.
5. Bode H.W., Variable Qualizer, *Bell Syst. Tech. J.*, Vol. 17, No. 2, 1938, pp. 229–244.
6. Takasaki Y., Generalized Theory of Variable Equalizers, *Proc. Int. Symp. Circuit and Syst.*, 1979, pp. 146–149.
7. Takasaki Y., Simple Inductorless Automatic Line Equalizer for PCM Transmission Using New Variable Transfer Function, *IEEE Trans. Commun.*, Vol. COM-26, No. 5, May 1978, pp. 675–678.
8. Takasaki Y., and Y. Takahashi, DMI/Duobinary Transmission Experiment in a Chain of Fiber-Optic Repeaters, *Int. Conf. Commun. '83*, June 19–22, 1983, Boston, MA, pp. 100–104.
9. Widmer A.X., and P.A. Franaszek, A DC-Balanced, Partitioned-Block, 8B/10B Transmission Code, *IBM J. Res. Develop.*, Vol. 27, No. 5, September 1983, pp. 440–451.
10. Takasaki Y., Two-level AMI Line Coding Family for Optical Fibre Systems, *Int. J. Electronics*, Vol. 55, No. 1, July 1983, pp. 121–131.

Appendix A
Minimum-Phase Transfer Function

If $\exp[A(\omega) + jB(\omega)]$ is a minimum-phase transfer function, the relation between amplitude $A(\omega)$ and phase $B(\omega)$ can be expressed as

$$B(\omega_c) = \frac{1}{\pi} \int_0^\infty \frac{dA(\omega)}{d\omega} \log \left| \frac{\omega + \omega_c}{\omega - \omega_c} \right| d\omega \qquad (A.1)$$

By applying this relation to the amplitude characteristic shown in Figure A.1, we obtain

$$B(\omega_c) = \frac{k\omega_0}{\pi} [(x + 1) \log(x + 1) + (x - 1) \log|x - 1| - 2x \log x] \qquad (A.2)$$

where $x = \omega_c/\omega_0$.

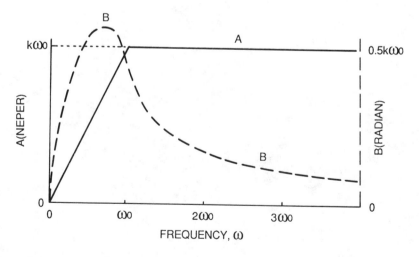

Figure A.1 Minimum-phase relation between amplitude $A(\omega)$ and phase $B(\omega)$ characteristics.

We can use the amplitude element A in Figure A.1 for approximating a given amplitude characteristic as shown in Figure A.2. The required number of $x \log x$ computations in calculating phase $\phi(\omega_i)(i = 1{-}m)$ by applying (A.2) for respective $A_j(\omega)(j = 1{-}n)$ is $3mn$. We can easily see that the number of computations can be reduced to $m + n$ when ω_i can be expressed as $\omega_i = i\Delta\omega(i = 1{-}m)$. This method is called the "fast minimum-phase transform" (FMT).

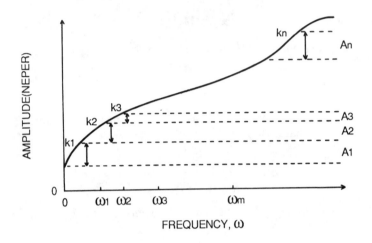

Figure A.2 Approximation of a given amplitude characteristic based on curve A in Figure A.1.

Appendix B
Bessel Filtering

Component values of Bessel filters shown in Figure B.1 (a) are listed in Table B.1. Approximating Nyquist's shaping for NRZ input pulse is assumed. Therefore, attenuation at $f_0/2$ is chosen to be 2 dB in Figure B.1(a) (see Problem 3.2). Amplitude, normalized delay, and output waveforms are shown in Figures B.1(a), (b) and (c), respectively, taking order of filter n as a parameter ($n = 1–11$). (See Figure B.2 for the case of RZ pulse.)

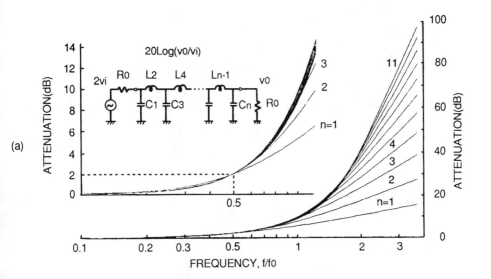

Figure B.1 Responses of Bessel filters for NRZ pulse: (a) attenuations in frequency domain; (b) delays in frequency domain; and (c) NRZ pulse response.

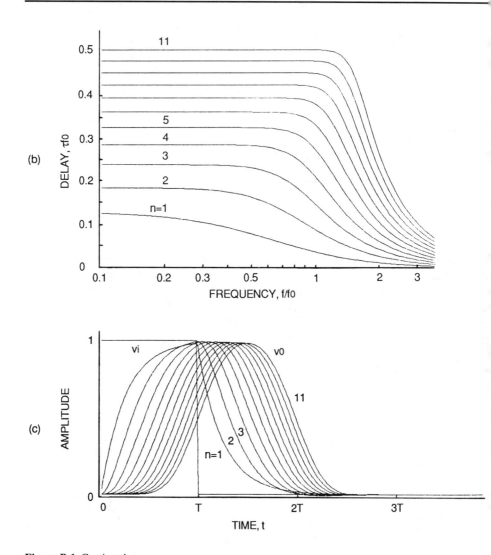

Figure B.1 Continued.

Figure B.2 Responses of Bessel filters for RZ pulse: (a) attenuations in frequency domain; (b) delays in frequency domain; and (c) RZ pulse response.

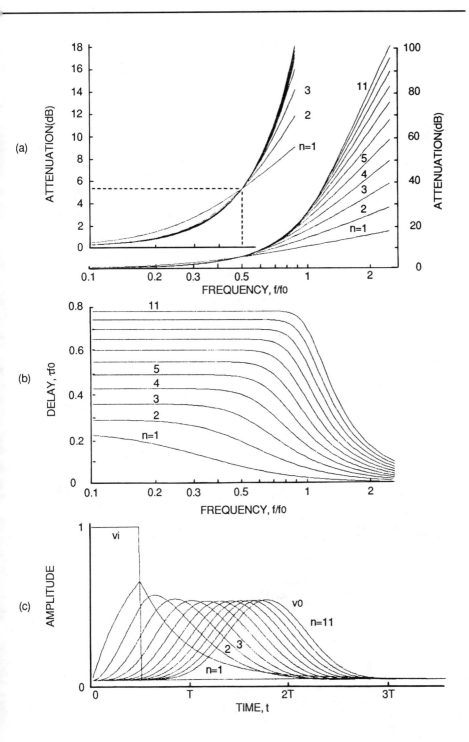

Table B.1
Component Values for Bessel Filters in Fig. B.1a

n	C_1	ℓ_2	C_3	ℓ_4	C_5	ℓ_6	C_7	ℓ_8	C_9	ℓ_{10}	C_{11}
1											.251
2										.286	.0767
3									.297	.131	.0455
4								.0494	.110	.343	.0671
5							.0528	.0934	.344	.124	.0369
6						.0374	.112	.348	.109	.0850	.0319
7					.0384	.101	.348	.120	.0949	.0645	.0222
8				.0224	.0629	.0913	.106	.347	.121	.0728	.0249
9			.0227	.0612	.0889	.0931	.343	.134	.0892	.0539	.0180
10		.0181	.0531	.0858	.123	.346	.103	.0903	.0735	.0483	.0168
11	.0182	.0522	.0831	.112	.347	.112	.0911	.0791	.0614	.0387	.0132

$$C_i = c_i \,/\, (R_o\, f_o)\ [F], \quad L_i = \ell_i\, R_o \,/\, f_o\ [H]$$

Appendix C
Simple Phase (Delay) Equalizer

A delay distortion is caused by a fairly sharp cut-off at the band edge. Based on this fact, a low-pass transmission characteristic $H(\omega)$ is presumed for the high-frequency cut-off as illustrated in Figure C.1(a). That is

$$H(\omega) = \exp[-\alpha(\omega) - j\theta(\omega)] \tag{C.1}$$

where $\alpha(\omega)$ is described in terms of a pulse repetition frequency $f_0(=\omega_0/2\pi)$ as

$$\alpha(\omega) \begin{cases} = 0, & |\omega| \leq \omega_0 \\ > 0, & |\omega| \geq \omega_0 \end{cases} \tag{C.2}$$

Phase $\theta(\omega)$ is related to amplitude $\alpha(\omega)$ by the Hilbert transform as

$$\theta(\omega) = (\omega/\pi) \int_{-\infty}^{\infty} \{\alpha(\omega_I)/(\omega_I^2 - \omega^2)\} \, d\omega_I \tag{C.3}$$

Delay $d\theta(\omega)/d\omega$ is described as

$$d\theta(\omega)/d\omega = (1/\pi) \int_{-\infty}^{\infty} \{\alpha(\omega_I)/(\omega_I^2 - \omega^2)\} \, d\omega_I$$
$$+ (2\omega^2/\pi) \int_{-\infty}^{\infty} \{\alpha(\omega_I)/(\omega_I^2 - \omega^2)^2\} \, d\omega_I \tag{C.4}$$

By using (C.2), we obtain

$$d\theta(\omega)/d\omega = (2/\pi) \int_{\omega_0}^{\infty} \{\alpha(\omega_I)/(\omega_I^2 - \omega^2)\} \, d\omega_I$$
$$+ (4\omega^2/\pi) \int_{\omega_0}^{\infty} \{\alpha(\omega_I)/(\omega_I^2 - \omega^2)^2\} \, d\omega_I \tag{C.5}$$

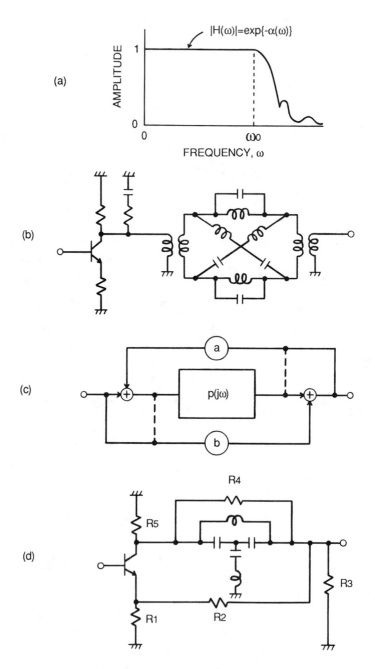

Figure C.1 Design of phase (delay) equalizer: (a) definition of high-frequency cut-off by $H(\omega)$; (b) example of passive delay equalizer; (c) principle of minimally active delay equalization; and (d) realization of minimally active delay equalizer.

Thus it is seen that

$$[d\theta(\omega)/d\omega]\omega = \omega_1 \geq [d\theta(\omega)/d\omega]\omega = \omega_2 \geq 0 \qquad (C.6)$$

for $\omega_0 \geq \omega_1 \geq \omega_2 \geq 0$. Therefore, we can conclude that delay caused due to the cut-off band ($\omega > \omega_0$) characteristics is monotone within the transmission band ($0 \leq \omega \leq \omega_0$).

This type of delay distortion can be compensated for by using delay equalizers, shown in Figure C.1(b). A second-order balanced type all-pass network is used in the figure. An unbalanced type all-pass network, which is desirable from the standpoint of simple implementation, cannot, however, compensate for such monotone increasing delay characteristics as given by (C.5). We have solved this problem by using a minimally active network illustrated in Figure C.1(c). Coefficients of feedback and feed-forward paths are found to be 0.2–0.3 in some typical examples. A more concrete realization is shown in Figure C.1(d). The transfer function of this network is given by

$$P(j\omega) = -G\{(1 - B) + B \cosh \theta + C \sinh \theta\}$$
$$/\{B' + (1 - B') \cosh \theta + F \sinh \theta\} \qquad (C.7)$$

where

$$B = -R_1/R_2 \qquad (C.8)$$

$$B' = B(2R_2 + R_4)/R_4 \qquad (C.9)$$

$$C = Z_0(R_4^{-1} + B/R_5) \qquad (C.10)$$

$$F = G\{(Z_0R_1/R_5)(1/R_3 + 1/R_2) + R_1/Z_0\} + Z_0/R_4 \qquad (C.11)$$

$$G = R_3R_5/R_1(R_3 + R_5) \qquad (C.12)$$

and θ and Z_0 are the image propagation constant and the image impedance of an all-pass network, respectively. There are four kinds of all-pass functions described by (C.7), depending on relations between B–G. A practical all-pass function is obtained by substituting the following relations into (C.7). That is

$$B = B' \qquad (C.13)$$

$$(1 - B)^2 + C^2 = B^2 + F^2 \qquad (C.14)$$

Appendix D
Generalized Expression for Variable Transfer Functions

An ideal variable transfer function can be expressed as $y_0(f)^u$, where $y_0(f)$ is a standard cable response and u corresponds to a parameter such as normalized cable length. It is assumed that $-1 \leq u \leq 1$. Since direct realization of $y_0(f)^u$ is difficult, an approximate realization $v(x, y_0(f))$ was proposed by Bode. Here, x is the value of a variable element such as a resister. Let us first study how u can be expressed as a function of x. The relation between v and y_0^u can be expressed as

$$v[u(x), y_0(f)] = y_0(f)^{u(x)} E[u(x), y_0(f)] \qquad \text{(D.1)}$$

where $E(u(x), y_0(f))$ is an error term. It is convenient to define u in terms of the following expansion:

$$\log(v(u, y_0)) = u \log y_0 + g_2(u)(\log y_0)^2 + \ldots \qquad \text{(D.2)}$$

Differentiating (D.2) in terms of y_0 and, after some manipulation, we obtain

$$u = [(y_0/v)(dv/dy_0)] \, y_0 = 1 = [(dv/dy_0)] \, y_0 = 1 \qquad \text{(D.3)}$$

since $v = 1$ when $y_0 = 1$, as seen in (D.2).

Some types of variable transfer functions can be generalized by using polynomial functions:

$$v(x, y_0) = z_n(y_0)x^n + z_{n-1}(y_0)x^{n-1} + \ldots + z_1(y_0)x + z_0(y_0) \qquad \text{(D.4)}$$

It is easily seen that $u(x)$ can be expressed as

$$u(x) = \hat{z}_n x^n + \hat{z}_{n-1} x^{n-1} + \ldots + \hat{z}_1 x + \hat{z}_0 \qquad \text{(D.5)}$$

where $\hat{z}_i \triangleq [dz_i/dy_0] \, y_0 = 1$

Other types of variable transfer functions can be generalized in terms of rational functions:

$$v(x, y_0) = \frac{z_n(y_0)x^n + z_{n-1}(y_0)x^{n-1} + \ldots + z_1(y_0)x + z_0(y_0)}{y_n(y_0)x^n + y_{n-1}(y_0)x^{n-1} + \ldots + y_1(y_0)x + 1} \tag{D.6}$$

The expression for u can be given by

$$u(x) = \frac{(\hat{z}_n - \hat{y}_n)x^n + (\hat{z}_{n-1} - \hat{y}_{n-1})x_{n-1} + \ldots + (\hat{z}_1 - \hat{y}_1)x + \hat{z}_0}{\tilde{y}_n x^n + y_{n-1}x^{n-1} + \ldots + \tilde{y}_1 x + 1} \tag{D.7}$$

where

$$\hat{y}_i \underline{\Delta} [dy_i/dy_0] \, y_0 = 1 \quad \text{and} \quad \tilde{y}_i \underline{\Delta} [y_i] \, y_0 = 1$$

We can define various variable transfer functions based on these expressions, as studied in Section 3.2.

Solutions

S.1 SOLUTIONS TO PROBLEMS IN CHAPTER 1

Solution 1.1 Although the costs of optical fibers are becoming comparable to those of metallic cables, optical transmitters and receivers, as well as such optical components as connectors and splicers, remain expensive. In trunk lines, expensive optical transmission equipment can be shared among many subscribers with the help of multiplexing, which is not the case with local systems.

Solution 1.2 We can use the worst-case eye diagram shown in Figure 1.3 for the conversion. See also Figure 8.1.

Solution 1.3 Poor dc balance and longer pulse-run length result in amplitude crosshair degradations due to intersymbol interference (ISI) caused by low-frequency cut-off. Longer pulse-run length can also cause larger jitter in the extracted clock component to lead to time crosshair degradations. (See Section 2.2.)

Solution 1.4 Near-end crosstalk (NEXT) is caused by signals that propagate to the opposite direction in a neighboring cable. Conversely, far-end crosstalk (FEXT) is caused by signals that propagate in the same direction.

Solution 1.5 Cut-off slope becomes steeper as bandwidth is made narrower. Therefore, larger phase distortion results (see Appendix A).

Solution 1.6 In wireless systems, the use of a smaller roll-off factor is very efficient in reducing interference from adjacent channels. On the other hand, the problem of interference from adjacent channels is not so important with metallic and optical fiber systems, because frequency division multiplexing is not usually used in such systems.

Solution 1.7 The cut-off slope of the raised cosine filter approaches infinity at pulse repetition frequency f_0. Therefore, extremely large (or infinite) phase distortion results when the minimum phase criterion is applied (see Appendix A).

Solution 1.8 Physically realizable waveforms usually obey Gibby-Smith criterion. Such waveforms assume narrower pulsewidth than those that obey Nyquist's criterion. An echo appearing at the center of a timeslot minimizes ISI: check the waveform with $n = 11$ in Figure 1.7(d).

Solution 1.9 Note that the sum of the echo amplitudes at T and $2T$ in Figure 1.8(a) is nearly equal to 0.05. Also note that the half-amplitude of the pulse waveform is 0.5. Therefore, the amount of ISI is 10%.

Solution 1.10 Pattern functions for patterns (b) and (d) sample the waveform function at $f_0/2$ and $f_0/4$, respectively. Therefore, the amount of jitter is around 4ε. In the same way, patterns (b) and (c) cause jitter of around 2ε.

Solution 1.11 Note that only lower-frequency components of jitter can accumulate. Type A jitter comprises only high-frequency components as we shall study in Section 6.1.

Solution 1.12 Note that the rise and fall times of timing jitter waveforms in Figure 1.18(a) are equal. However, we have different rise and fall times in actual systems, especially for a quasistatic $1/8-8/8$ pattern, as we shall study in Section 7.2. That is, when the pattern is changed from 8/8 to 1/8, the change of phases in the clock component is slower, compared to the case where the pattern is changed from 1/8 to 8/8. In such a case, the negative spike of the alignment jitter is shorter than the positive spike.

Solution 1.13 The phase of the clock component is forced to assume a particular value at the moment a pulse is applied to a narrow-band filter for clock extraction. This phenomenon is prominent when Q is low and the pulse pattern is sparse. This is the jitter-fine structure we will study in Section 5.1.

Solution 1.14 The use of 4B3T code requires scrambling of the pulse stream to avoid false word synchronization.

Solution 1.15 Note that the twisted-pair cable also assumes a square root of f response. The loss of cable increases by a factor of 10 when the equivalent transmission frequency (or Nyquist frequency: $f_0/2$) is increased by a factor of 100. Therefore, repeater spacing is reduced by a factor of 10.

S.2 SOLUTIONS TO PROBLEMS IN CHAPTER 2

Solution 2.1 Note that RZ and NRZ pulses have spectral nulls at $2f_0$ and f_0, respectively. Therefore $x = 2\ \pi f/f_0$ and $\pi f/f_0$, respectively.

Solution 2.2 Assume that each block in an NRZ pulse stream with unit amplitude has pattern "11100." Applying this pulse stream to ac coupling results in positive and negative amplitudes of 0.4 and 0.6, respectively. Therefore, worst-case ISI is 20%.

Solution 2.3 Coding efficiency is the same with both coding schemes. Although ZS is not required with PST, the use of a scrambler is a prerequisite for avoiding false word synchronization. However, AMI can choose whether to use ZS or scrambling.

Solution 2.4 Use sequences "1010 ..." and "11001100 ..." Show that these sequences are precoded to result in sequences "10001000 ..." and "1111000011110000 ... ," respectively. Next, show that we can obtain sequences "10-1010-10 ..." and "1100-1-1001100-1-100 ... ," respectively, after applying modified duobinary filtering. Finally, show that these two sequences have different peak amplitudes.

Solution 2.5 Delay the waveform in Figure 2.10(d) by $2T$. Subtract the delayed waveform from the original one. Compare the obtained waveform with the one in Figure 2.10(c), keeping in mind the description in Section 2.5.3 that T is halved.

S.3 SOLUTIONS TO PROBLEMS IN CHAPTER 3

Solution 3.1 Normalize the curve labelled "Gaussian" in Figure 3.2(a) to assume an amplitude of 0.5 at $f_0/2$, and compare with Figure 3.1(b).

Solution 3.2 Apply $x = \pi f/f_0$ in Figure 3.6(a) to obtain a response of 0.64 at $f_0/2$. Therefore, $20 \log(0.5/0.64) \simeq -2$ dB.

Solution 3.3 Calculate $\sqrt{j\omega} = \alpha + j\beta$ to obtain $\alpha = \beta = \sqrt{\omega/2}$. Then we obtain

$$20 \log|C(j\omega)| = 20 \log[\exp(-B \sqrt{\omega/2})], \text{ (for } A = 1)$$
$$= -20B \,(\log e) \sqrt{\pi f} \,(\text{dB})$$

Solution 3.4 Assume that an x element that can attain $0.05 \le x \le 20$ is available. From (3.14), we obtain $0.9 \gtrsim u \gtrsim -0.9$. Then we can find that $L_v = 20$ dB is obtained by designing $y_0(f_e) = 0.077$.

Solution 3.5 See Figure 8.8(a).

Solution 3.6 Let us assume zero-forcing at (x_1, u_1), (x_2, u_2), (x_3, u_3). Solve $v(x_i) = y_0^{u_i}$ for $i = 1, 2, 3$ to obtain z_0, z_1 and y_1. Then we can derive $\hat{z}_1$, $\hat{y}_1$, $\hat{z}_0$ and $\bar{y}_1$ by using the relations described in Appendix D. Finally, we can eliminate variable x from (3.16) by using relation (3.17) to find that v can be expressed without the use of x_i ($i = 1, 2, 3$).

Let us study (3.18) as a simpler example. By following the procedure mentioned above, we can derive

$$v = \frac{(u_1 - u) + (u_1 + u)y_0^{u_1}}{(u_1 + u) + (u_1 - u)y_0^{u_1}}$$

That is, v is expressed without the use of x_1.

Solution 3.7 We can express a waveform $g(t)$ with a main response $h(t)$ and its echoes by using their Fourier transforms $G(j\omega)$ and $H(j\omega)$, respectively, as

$$G(j\omega) = H(j\omega)[\varepsilon_1 \exp(j\omega\tau_1) + 1 + \varepsilon_2 \exp(-j\omega\tau_2)]$$

For $\varepsilon_1 = \varepsilon_2 = \varepsilon$ and $\tau_1 = \tau_2 = \tau$ we obtain

$$G(j\omega) = H(j\omega)[1 + 2\varepsilon \cos(\omega\tau)]$$

to know that Figure 3.19(a) describes exact relations. For $-\varepsilon_1 = \varepsilon_2 = \varepsilon$ and $\tau_1 = \tau_2 = \tau$

$$G(j\omega) = H(j\omega)[1 - j2\varepsilon \sin(\omega\tau)]$$

Therefore, Figure 3.19(b) represents approximate relations for $\varepsilon \ll 1$. If $\varepsilon_1 = 0$, $\varepsilon_2 = \varepsilon$, $\tau_2 = \tau$,

$$G(j\omega) = H(j\omega)[1 + \varepsilon \exp(-j\omega\tau)]$$
$$= H(j\omega)[1 + \varepsilon \cos(\omega\tau) - j\varepsilon \sin(\omega\tau)]$$

We can see that Figure 3.19(c) is obtained by averaging Figure 3.19(a) and (b).

Solution 3.8 Note that the sum of the amplitudes at $-T/2$ and $T/2$ of the raised cosine waveform in Figure 3.2(b) is almost 1. Therefore, the worst-case ISI contributed by an echo with $\varepsilon = 0.05$ is ± 0.05. Since the half-eye opening is 0.5, ISI is 10%. Also note that smaller ISI is encountered when we use a Gaussian waveform.

S.4 SOLUTIONS TO PROBLEMS IN CHAPTER 4

Solution 4.1 See the raised cosine and Gaussian waveforms in Figure 3.2(b), and recall that they follow Nyquist's and the Gibby-Smith criteria, respectively. We can clearly see that the horizontal eye opening for the Gaussian waveform is narrower than that for the raised cosine waveform.

Solution 4.2 As should be obvious, resistance R connected at the output of an $n:1$ transformer shows equivalent resistance of n^2R at the input to the transformer. Next, let us assume that L is disconnected from c_1 and c_2. We can see that the voltage that appears at the connection of c_1 and c_2 is $(c_1/c_1 + c_2)v = kv$ for sufficiently small r. The impedance seen from this connecting point is $1/[j\omega c_2 + 1/(r + 1/j\omega c_1)] \approx [1/j\omega(c_1 + c_2)] + (c_1/c_1 + c_2)^2 r$, for sufficiently small r.

We can see that the combined use of c_1 and c_2 contributes to reduce an equivalent value of r by a factor of k^2, and also that the $n : 1$ transformer increases an equivalent value of R by a factor of n^2. This leads to improved quality factor Q.
The parallel connection of L and n^2R can be expressed as

$$j\omega Ln^2R/(j\omega L + n^2R) \approx j\omega L + (\omega L)^2/(n^2R)$$

for $\omega L \ll n^2R$. Therefore,

$$Q \approx \omega_0 L/r_e$$

where

$$r_e = k^2r + (\omega L)^2/(n^2R)$$
$$\omega_0 L = 1/[\omega_0(c_1 + c_2)]$$

Solution 4.3 The input waveform with a frequency f_i is multiplied with the output waveform with a frequency f_o to generate a high frequency component $(f_i + f_o)$ and a low frequency component $(f_i - f_o)$. The former is suppressed by the loop filter. If difference $f_i - f_o$ is in a pull-in range Δf_p, the phase of the lower-frequency component oscillates to converge, and after the difference enters a lock-in range Δf_L, the phase assumes a constant value after a single cycle of oscillation.

Solution 4.4 Full-wave rectified waveforms in Figures 4.5(b) and (b') comprise in-phase components (b), (b'), (c), and (c') as well as out-of-phase components (e) and (e'). The latter causes jitter. Note, however, that out-of-phase component (e) does not cause jitter because it has a symmetrical waveform and is located at the center of a timeslot.

Solution 4.5 See Figure 3.3.

S.5 SOLUTIONS TO PROBLEMS IN CHAPTER 5

Solution 5.1 Note that only low-frequency components of jitter can accumulate in a chain of repeaters. Waveforms in Figures 5.3(c) and 5.4(c) can include low-frequency components when the change of patterns is very slow (type B). However, other waveforms in Figures 5.3 and 5.4 can incorporate only high-frequency components, no matter how slow the change of digital pattern might be (type A).

Solution 5.2 Note that physically realizable waveforms usually obey the Gibby-Smith criterion; also see Solution 1.8.

Solution 5.3 See Figure 5.6(b). Curve #3 is too complicated and curve #1 is too simple to be encountered in a practical design. Therefore, curve #2 is the most probable. See also Figure 5.6(c). Curve #1 is very similar in shape to phases of waveform functions shown in Figure 5.5($a2'$), if the echo amplitude is inverted. Therefore, this suggests that echo #1 is predominant in Bessel-type waveform shaping.

Solution 5.4 In an approximate sense, we choose f_b so that

$$H_{BM}(f_0/2)/H_{BM}(0) = \begin{cases} 0.56 & \text{(for RZ pulse)} \\ 0.78 & \text{(for NRZ pulse)} \end{cases}$$

See Figure 3.6 and Solution 3.2.

Solution 5.5 See Figure 5.5($a2'$) and check the curve for $n = 5$ with curves in Figure 4.6. Note that the phase of the clock component can be given by the inner product of waveform functions in Figure 5.5 and pattern functions in Figure 4.6. AMI and straight binary sequences with the 1/2 pattern cause static jitter of around 20° and 0°, respectively. However, AMI and straight binary with 8/8 pattern cause static jitter of around 0° and 50°, respectively. If we choose the 1/2 pattern as a reference, 8/8 pattern leads to relative jitter of around −20° and 50° for AMI and straight binary sequences, respectively.

S.6 SOLUTIONS TO PROBLEMS IN CHAPTER 6

Solution 6.1 Note that the timing filter is equivalent to a low-pass filter as far as phase modulation is concerned. Also note that type A jitter does not include low-frequency components. Therefore, the attenuation of high-frequency components in a repeater chain is so large that the amount of accumulated type A jitter converges to a constant as the number of repeaters increases.

Solution 6.2 Note that type B jitter is caused by echo #2 in Figure 5.6(a) with $\varepsilon = 0.05$. Therefore, we can evaluate the amount of jitter caused by an echo, by referring to Figures 5.6(c) and 4.6. The results are shown in Figure 5.7. It should be self-explanatory from this figure that 1/8–8/8 pattern generates smaller jitter than 1/2–8/8.

Solution 6.3 Note that timing information is carried by pulses indicating digital marks. The density of digital marks in 1/8 pattern is smaller than in 1/2 pattern. Therefore, mistuning causes more susceptible 1/8 pattern to result in larger phase shifts.

Solution 6.4 Note that jitter introduced at the second repeater can be expressed as $\theta(t - MT)$. This can be understood by recalling the fact that jitter injected at each repeater is pattern-dependent. Similarly, we can understand that jitter injected at the third repeater can be expressed as $\theta(t - 2MT)$. We can extrapolate these to obtain an expression for the Nth repeater.

Solution 6.5 By resorting to qualitative analysis as shown in Figure 6.7(a), we can show that the spike-shaped waveform with negative polarity just in front of the leading edge of the jitter waveform, moves to just after the leading edge for a negative M value, as the positive spike moves to just after the trailing edge without changing the polarity. Therefore, we can understand that a negative M value tends to discourage extra-accumulation (see also the curve labelled -30 in Figure 6.7(c)).

Solution 6.6 T_p can be relatively small in a system where scrambling is adopted. However, for a system that incorporates the transmission of still pictures, T_p can become on the order of a million.

Solution 6.7 Let us first define transitional jitter, which is the worst-case jitter because it includes both worst-case static jitter and worst-case transitional jitter. This type of jitter is encountered by the use of a sufficiently long pulse sequence with a pattern combination that causes worst-case static jitter. Such a sequence causes square-wave jitter of very long period, with transitional jitter at the leading and trailing edges of the square wave. We call it "quasistatic jitter" when the length of the sequence is defined in accordance with the requirements and specifications of practical system design, and ceases to be sufficiently long.

Now the difference between static jitter, quasistatic jitter and transitional jitter should be self-explanatory. That is, transitional jitter is included in quasistatic jitter when the length of a pulse sequence is sufficiently long, but transitional jitter at the leading and trailing edges starts to interfere as the length of the sequence, or the spacing between the two edges, becomes shorter.

S.7 SOLUTIONS TO PROBLEMS IN CHAPTER 7

Solution 7.1 Note that jitter in Figure 7.2 is caused by echo #2 in Figure 5.6(a) with $\varepsilon = 0.05$. Therefore, we can evaluate the amount of jitter caused by an echo, by referring to Figures 5.6(c) and 4.6. The results are shown in Figure 5.7. It should be self-explanatory from this figure that the $1/2$–$8/8$ pattern generates a larger amount of jitter than other patterns.

Solution 7.2 Note that timing information is carried by pulses indicating digital marks. The density of digital marks in $8/8$ pattern is larger than in any other patterns. Therefore, the time required for pulling in the clock phase with $8/8$ pattern is shorter than with any other patterns.

Solution 7.3 Note that timing information is carried by pulses indicating digital marks. Also note that the time required for pulling in the clock phase depends on the difference of mark densities between the two patterns under consideration. Therefore, jitter waveforms for pattern 1/8–8/8 have slower rise but faster fall times, compared to the 1/2–8/8 pattern.

Solution 7.4 Note that M-dependent peaks of curves in Figure 7.3 move to the right as Q becomes larger (that is, as rise and fall times become longer). Also note that rise time is much longer than fall time with waveforms of 1/8–8/8 pattern in Figure 7.2. Therefore, the negative peaks that correspond to the leading edge with slower rise times appear to the right of the positive peaks.

Solution 7.5 Note that jitter has the same waveform at the input and output of a repeater if N is larger than 1 (input jitter is zero for $N = 1$), except for the delay of jitter waveform as far as fine structures are concerned. This is because the fine structure in incoming jitter is filtered out by the low-pass characteristic of the timing filter and only the fine structure of injected jitter appears at the output of the repeater. Therefore, we can see that the fine structure in alignment jitter is given by the difference between such waveforms having the same shape, but different delay, as the one shown in Figure 7.5(d) for $N = 1$. As a consequence, the amplitude of fine structures of alignment jitter is doubled at the second repeater, and then stays constant for N larger than 2.

S.8 SOLUTIONS TO PROBLEMS IN CHAPTER 8

Solution 8.1 We can obtain the curve in Figure 8.1 by using the eye diagram shown in Figure 1.3 (solid line). We have to be careful in converting the degradation of time crosshair budget to amplitude crosshair budget based on the hybrid design criterion. In Table 8.2, for example, we first have 15% degradation, for a time crosshair degradation of 64°, and then have a degradation of 19% for a time crosshair degradation of 71.5°. Therefore, rms degradation is 4% ($=19 - 15$). We cannot obtain a correct conversion if we use the curve at 7.5° to obtain an rms degradation of 1%.

Solution 8.2 The cost of an LED multimode fiber system for 1 km transmission with a 100 m repeater spacing is $4,640 ($3,640 for 10 repeaters and $1,000 for 1 km of optical fiber). However, a metallic cable system costs $2480 ($1730 for 10 repeaters and $750 for 1 km of coaxial cable). Therefore, the metallic cable system remains less expensive. Further cost reduction can be achieved if we can use twisted-pair cables instead of coaxial cables. Note that the cost of cable installation, which would adversely affect optical system economy, is not included in the above calculation.

Solution 8.3 By comparing Figures 8.7(b) and 4.6, we can understand that a sequence, as shown in Figure 4.6(a), that can cause worst-case jitter (that is, the pattern function confined around the zero frequency), is not encountered by a pulse train that incorporates the AMI format. Therefore, we can enjoy much smaller jitter than the worst-case prediction in terms of waveform function when we employ the AMI format.

Solution 8.4 We can see that the pulses indicated by an arrow are centered at $\pm T/2$ from the center of a timeslot. Therefore, the clock component extracted from such a pulse has a phase difference of 180° with that extracted from normal pulses. As a consequence, pulses indicated by an arrow tend to cancel the extracted clock component.

Solution 8.5 We can see from Figure 8.14(f′) that noise with doubled amplitude would have to be applied for regenerating $0(+)$ or $0(-)$ as -1 or $+1$, respectively, compared to cases where 1 or -1 is erroneously regenerated as 0, and *vice versa*. We can show that the possibility of undergoing such noise is negligibly small when pulse error rate is sufficiently small.

Solution 8.6 We can use the pulse that indicates the detection of eight successive zeros to inhibit ϕ_2 after delaying it by 4 and 7 timeslots.

INDEX

The Artech House Telecommunications Library

Vinton G. Cerf, *Series Editor*

A Bibliography of Telecommunications and Socio-Economic Development by Heather E. Hudson

Advances in Computer Systems Security: 3 volume set, Rein Turn, ed.

Advances in Fiber Optics Communications, Henry F. Taylor, ed.

Broadband LAN Technology by Gary Y. Kim

Codes for Error Control and Synchronization by Djimitri Wiggert

Communication Satellites in the Geostationary Orbit by Donald M. Jansky and Michel C. Jeruchim

Current Advances in LANs, MANs, and ISDN, B.G. Kim, ed.

Design and Prospects for the ISDN by G. DICENET

Digital Cellular Radio by George Calhoun

Digital Image Signal Processing by Friedrich Wahl

Digital Signal Processing by Murat Kunt

Digital Switching Control Architectures by Giuseppe Fantauzzi

Digital Transmission Design and Jitter Analysis by Yoshitaka Takasaki

Disaster Recovery Planning for Telecommunications by Leo A. Wrobel

E-Mail by Stephen A. Caswell

Expert Systems Applications in Integrated Network Management, E.C. Ericson, L.T. Ericson, and D. Minoli, eds.

Handbook of Satellite Telecommunications and Broadcasting, L. Ya. Kantor, ed.

Innovations in Internetworking, Craig Partridge, ed.

Integrated Services Digital Networks by Anthony M. Rutkowski